Stata Tips

Volume II: Tips 120–152

Fourth Edition

Stata Tips

Volume II: Tips 120–152

Fourth Edition

NICHOLAS J. COX, Editor
Durham University
Department of Geography

A Stata Press Publication
StataCorp LLC
College Station, Texas

Second edition 2009
Third edition 2014
Fourth edition 2024

Published by Stata Press, 4905 Lakeway Drive, College Station, Texas 77845
Typeset in LaTeX 2ε
Printed in the United States of America
10 9 8 7 6 5 4 3 2 1

Print ISBN-10: 1-59718-405-5 (volumes I and II)
Print ISBN-10: 1-59718-407-1 (volume I)
Print ISBN-10: 1-59718-409-8 (volume II)
Print ISBN-13: 978-1-59718-405-2 (volumes I and II)
Print ISBN-13: 978-1-59718-407-6 (volume I)
Print ISBN-13: 978-1-59718-409-0 (volume II)
ePub ISBN-10: 1-59718-406-3 (volumes I and II)
ePub ISBN-10: 1-59718-408-X (volumes I)
ePub ISBN-10: 1-59718-410-1 (volumes II)
ePub ISBN-13: 978-1-59718-406-9 (volumes I and II)
ePub ISBN-13: 978-1-59718-408-3 (volumes I)
ePub ISBN-13: 978-1-59718-410-6 (volumes II)

Library of Congress Control Number: 2023948738

Contents

Subject table of contents

General

Data management

Graphics

Programming

Statistics

Editor's preface

The book you are reading reprints 33 Stata Tips from the *Stata Journal* from 2014 to 2023, with thanks to their original authors. It is a sequel to *One Hundred Nineteen Stata Tips* from 2014, reissued together with this volume. The *Journal* began publishing tips in 2003, beginning with volume 3, number 4. The Editors are now pleased to reprint this selection in this book. Among past and present Editors, Nicholas J. Cox has overseen the production of these Tips from the beginning, with continued support and encouragement from H. Joseph Newton and Stephen P. Jenkins.

The *Stata Journal* publishes substantive and peer-reviewed articles ranging from reports of original work to tutorials on statistical methods and models implemented in Stata, and indeed on Stata itself. Other features include regular columns such as "Speaking Stata", book reviews, and announcements.

We are pleased by the external recognition that the *Journal* has achieved. The *Stata Journal* is indexed and abstracted by CompuMath Citation Index, Current Contents/Social and Behavioral Sciences, RePEc: Research Papers in Economics, Science Citation Index Expanded (also known as SciSearch), Scopus, and Social Sciences Citation Index.

But back to the Tips. There was little need for tips in the early days. Stata 1.0 was released in 1985. The original program had 44 commands, and its documentation totaled 175 pages. The current version, on the other hand, has hundreds if not thousands of commands—including an embedded matrix language called Mata—and Stata's official documentation now totals more than 18,000 pages. Beyond that, the user community has added several hundred more commands and many more pages explaining them or the official commands.

The pluses and the minuses of this growth are evident. As Stata expands, it is increasingly likely that users' needs can be met by available code. But at the same time, learning how to use Stata and even learning what is available become larger and larger tasks.

The Tips are intended to help. The ground rules for Stata Tips, as found in the original 2003 statement, are laid out as the next item in this book. We have violated one original rule in the letter, if not the spirit: some Stata Tips have been much longer than three pages. However, the intention of producing concise tips that are easy to pick up remains as it was.

The Tips grew from many discussions and postings on Statalist, at Stata conferences, meetings, and workshops, and elsewhere, which underscores a simple fact: Stata is now so big that it is easy to miss even simple features that can streamline and enhance your sessions with Stata. This applies not just to new users, who understandably may quake nervously before the manual mountain, but also to longtime users, who too are faced with a mass of new features in every release.

Tips have come from Stata users as well as from StataCorp employees. Many discuss new features of Stata, or features not documented fully or even at all. We hope that you enjoy the Stata Tips reprinted here and can share them with your fellow Stata users. If you have tips that you would like to write, or comments on the kinds of tips that are helpful, do get in touch with us, as we are eager to continue the series.

Among many complementary resources, and beyond the all-important help files and manual volumes, I want to flag two features of the StataCorp website, https://www.stata.com, namely, the FAQs ("Frequently asked questions on using Stata") and the Stata Blog, *Not Elsewhere Classified*. Both share the primary aims of alerting you to features of Stata and how to use them easily and effectively. They also include many contributions from the user community.

Nicholas J. Cox, Editor
October 2023

The Stata Journal (2003)
3, Number 4, p. 328 DOI: 10.1177/1536867X0400300402

Introducing Stata tips

As promised in our editorial in *Stata Journal* 3(2), 105–108 (2003), the *Stata Journal* is hereby starting a regular column of tips. Stata tips will be a series of concise notes about Stata commands, features, or tricks that you may not yet have encountered.

The examples in this issue should indicate the kinds of tips we will publish. What we most hope for is that readers are left feeling, "I wish I had known that earlier!" Beyond that, here are some more precise guidelines:

Content A tip will draw attention to useful details in Stata or in the use of Stata. We are especially keen to publish tips of practical value to a wide range of users. A tip could concern statistics, data management, graphics, or any other use of Stata. It may include advice on the user interface or about interacting with the operating system. Tips may explain pitfalls (do not do this) as well as positive features (do use this). Tips will not include plugs for user-written programs, however smart or useful.

Length Tips must be brief. A tip will take up at most three printed pages. Often a code example will explain just as much as a verbal discussion.

Authorship We welcome submissions of tips from readers. We also welcome suggestions of tips or of kinds of tips you would like to see, even if you do not feel that you are the person to write them. Naturally, we also welcome feedback on what has been published. An email to *editors@stata-journal.com* will reach us both.

H. Joseph Newton, Editor
Texas A&M University
jnewton@stat.tamu.edu

Nicholas J. Cox, Editor
University of Durham
n.j.cox@durham.ac.uk

2 DOI: 10.1177/1536867X1401400216

The Stata Journal (2014)
14, Number 2, pp. 449–450

Stata tip 120: Certifying subroutines

Maarten L. Buis
Wissenschaftszentrum Berlin für Sozialforschung (WZB)
Berlin, Germany
maarten.buis@wzb.eu

When you write your own program in Stata, it is good practice and useful to create and run a certification script. A certification script is a do-file that runs your program and compares the results either with a result that is known to be true or with results from a previous run (Gould 2001). It is also good practice and useful to divide your program into smaller subroutines; you can store these subroutines in the same ado-file. These subroutines will be visible only to other programs defined within the same ado-file; the only program that is visible to all other programs in Stata will be the first program defined in an ado-file. This can be useful for subroutines that make sense only within the context of the main program. For example, you may want to delegate the parsing of some complicated syntax element to a subroutine. Moreover, putting a subroutine inside the ado-file of the main program protects users against accidentally running that subroutine. This can be important when, for example, the subroutine changes the data, and the main program has various safeguards in place to ensure that this will not corrupt the user's data.

Sometimes, it is helpful to certify some subroutines in isolation. How would you do that if the subroutines are not visible outside the ado-file? You could copy the subroutine and store it in its own file, thus making the subroutine globally visible. However, as mentioned above, there can be good reasons why you would not want the subroutine to be globally visible in the final program.

Another solution is to `do` or `run` (see [R] **do**) the ado-file. This treats the ado-file as a regular do-file, which in this case defines only a set of programs. So after you `do` or `run` an ado-file, all of its subroutines will also be available. Now you can certify the subroutines from the file that will be released to the general public without having to copy and paste parts out of and into that file. This trick can also be useful when debugging a subroutine.

Consider the example ado-file below:

```
*! version 1.0.0 26Feb2014 MLB
program mainprog
version 13
args input
subprog `input'
di `"`s(output)'"'
end

program subprog, sclass
version 13
args input
sreturn local output `"do something smart with "`input'""'
end
```

If you store this file where Stata can see it (see [P] **sysdir**) or if you change the working directory to where this ado-file is stored (see [D] **cd**), the command `mainprog` works directly. However, if you try to call `subprog`, Stata will return an error.

```
. clear all

. mainprog "this"
do something smart with "this"

. subprog "this"
unrecognized command:  subprog
r(199);
```

We can look at the names of the programs stored in memory by using `program dir` (see [P] **program**), and we see that `subprog` exists but only as part of the `mainprog` command.

```
. program dir
  ado       232  mainprog.subprog
  ado       213  mainprog
  (output omitted)
```

If we `run` this ado-file first, then we can directly access both `mainprog` and `subprog`.

```
. clear all

. run mainprog.ado

. mainprog "this"
do something smart with "this"

. subprog "this"

. di `"`s(output)'"'
do something smart with "this"

. program dir
            232  subprog
            213  mainprog
  (output omitted)
```

Reference

Gould, W. 2001. Statistical software certification. *Stata Journal* 1: 29–50. https://doi.org/10.1177/1536867X0100100102.

4 DOI: 10.1177/1536867X1401400417

The Stata Journal (2014)
14, Number 4, pp. 991–996

Stata tip 121: Box plots side by side

Nicholas J. Cox
Department of Geography
Durham University
Durham, UK
n.j.cox@durham.ac.uk

Box plots are a standard plot type in statistical graphics and, as such, are popular with Stata users. The official Stata commands `graph box` and `graph hbox` are identical except that `graph box` draws box plots with the response (or outcome) scale on the vertical axis and `graph hbox` draws plots with the response scale on the horizontal axis. Contrary to the usual mathematical convention, the response axis is always regarded as the y axis for these commands so that options such as `ytitle()`, `ylabel()`, and `yscale()` always apply to the axis with the response variable. The manual entry [G-2] **graph box** gives much more detail and pertinent references. For a wider discussion of box plots, including how to draw box plots and related plots with `graph twoway`, see Cox (2009, 2013).

The greatest value of box plots is for comparing distributions of related variables or distributions of single variables for different groups of observations. This tip focuses on how and which data are plotted side by side. I explain the default appearance and structure of side-by-side box plots and how to tune or even to reverse that default.

To make this question concrete, we read in some data and then plot some graphs. As often happens, the code here is a cleaned-up version of what was done in preparing the tip, with afterthoughts and second guesses turned into anticipations of useful ideas.

```
. set scheme sj
. sysuse citytemp
. local title "Mean temperatures ({&degree}F)"
. label var tempjan "January"
. label var tempjul "July"
```

The `citytemp` dataset distributed with Stata contains temperature data for various U.S. cities. These are given in degrees Fahrenheit, a scale on which water freezes at 32° F and water boils at 212° F. Most countries of the world use the Celsius (formerly centigrade) scale ° C, for which water freezes at 0° C and boils at 100° C. The degree symbol can be shown as a text symbol, as explained in the help for `text`. If this functionality is not available in your Stata, you can use the trick explained in Cox (2004). One way or another, we create a local macro indicating units of measurement for use in later graphs. Because we plan to use a graph title explaining that we are showing temperatures, the month names suffice as variable labels.

Some examples of box plots are shown in figures 1 and 2. Figures 1a and 2a show vertical box plots for January and July temperatures in various regions of the United States, while figures 1b and 2b show corresponding horizontal box plots. The commands are as follows:

```
. graph box tempjan tempjuly, over(region) ytitle(`title´)
> ylabel(14 32 50 68 86, angle(h))
. graph hbox tempjan tempjuly, over(region) ytitle(`title´) ylabel(14(18)86)
. graph box tempjan tempjuly, by(region, rows(1) compact note(""))
> ytitle(`title´) ylabel(14(18)86, angle(h))
. graph hbox tempjan tempjuly, by(region, cols(1) compact note(""))
> ytitle(`title´) ylabel(14(18)86)
```

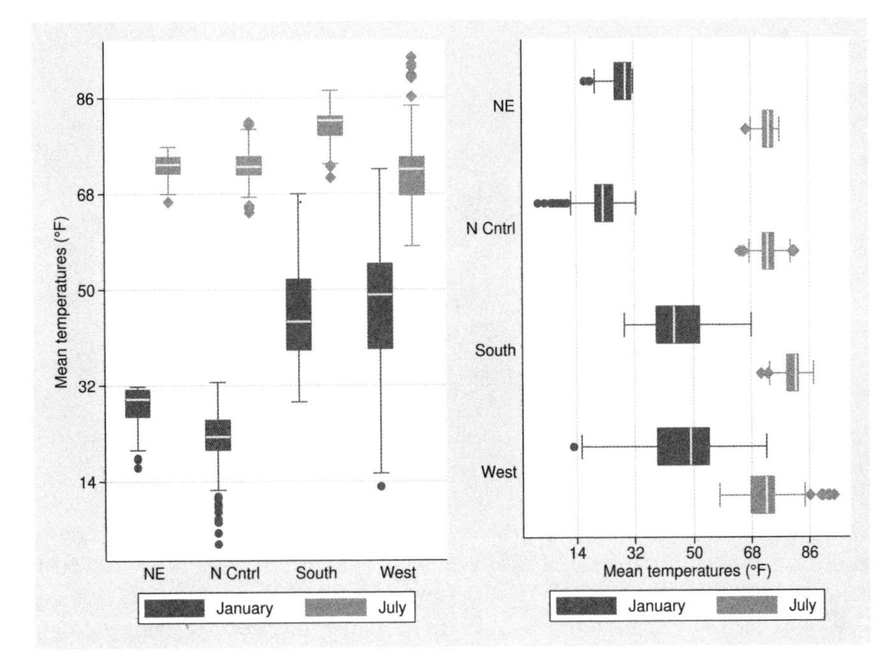

Figure 1. Box plots for January and July temperatures of various U.S. cities using the `over()` option to compare different regions

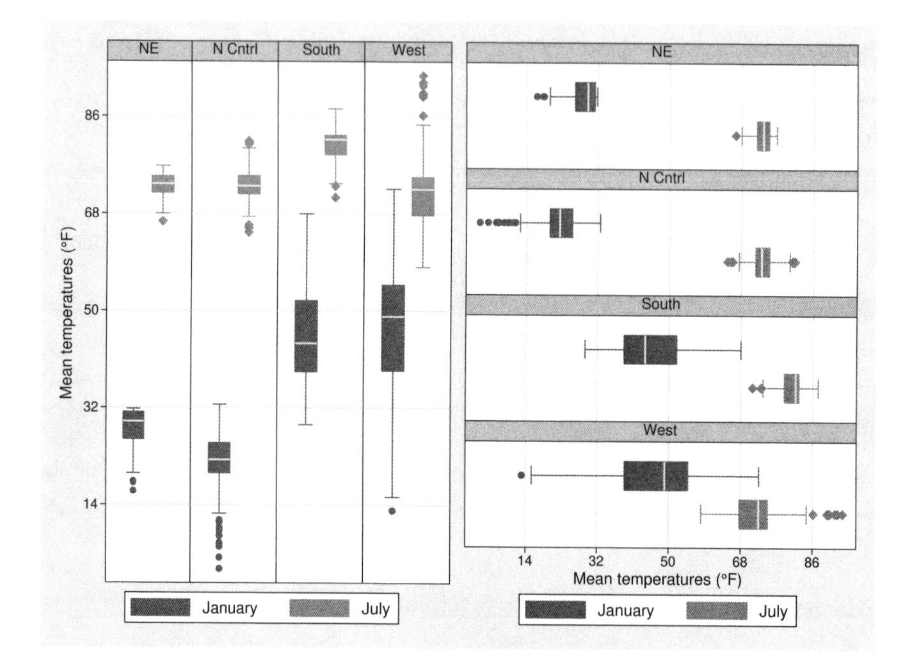

Figure 2. Box plots for January and July temperatures of various U.S. cities using the `by()` option to compare different regions

The axis labels 14(18)86 may seem a strange choice to U.S. readers, but 32° F is a key threshold, while differences of 18° F between labeled ticks match differences of 10° C. Figure 1 uses the `over()` option to compare different regions, while figure 2 uses the `by()` option to compare regions. In broad terms, the `by()` option is more flexible but produces more scaffolding. The scaffolding is sometimes helpful in indicating the subdivisions of the graph clearly but sometimes less helpful in that it may take up valuable space. Users aware of both syntaxes can make an informed choice.

What is less well known is that the `by()` option can be tuned so that results resemble those of the `over()` option. This trick may be applied more widely than just to box plots. Appropriate incantations tweak the position and appearance of the subtitles of the component graphs. It is convenient, but not essential, to define those incantations with local macros for repeated use in later commands. Note the clock notation for position, which places subtitles for vertical box plots at 12 o'clock and those for horizontal plots at 9 o'clock. Figure 3 shows the results.

```
. local incant1 subtitle(, position(12) ring(1) nobexpand bcolor(none)
> placement(n))
. local incant2 subtitle(, position(9) ring(1) nobexpand bcolor(none) placement(e))
. graph box tempjan tempjuly, by(region, rows(1) compact note(""))
> ytitle(`title´) ylabel(14(18)86, angle(h)) `incant1´
. graph hbox tempjan tempjuly, by(region, cols(1) compact note(""))
> ytitle(`title´) ylabel(14 32 50 68 86) `incant2´
```

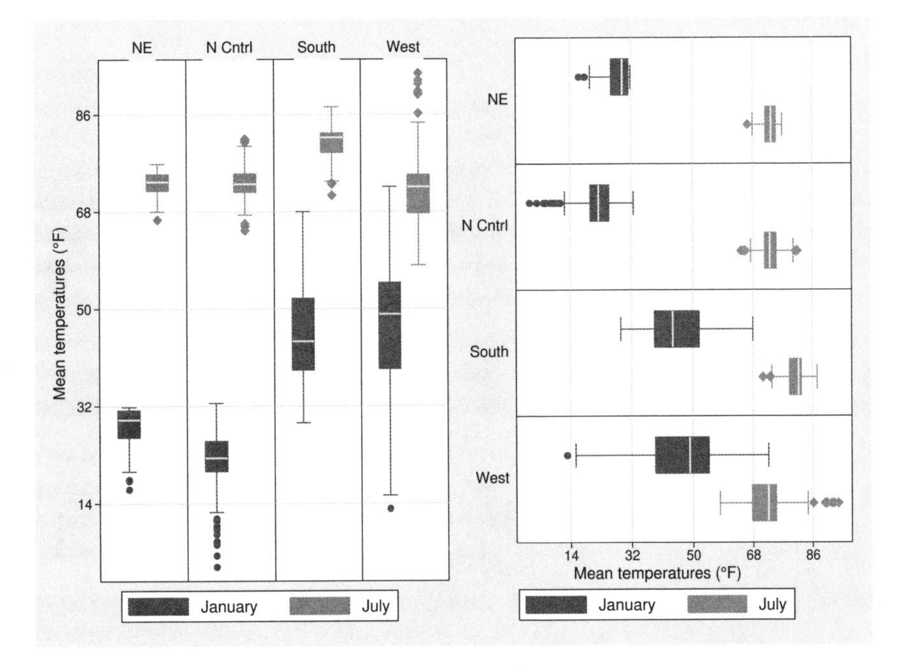

Figure 3. Box plots for January and July temperatures of various U.S. cities using the by() option to compare different regions, but with panel titles shown differently

Despite these minor variations, the design common to all the plots so far is that different variables are placed closest (on the inside, as it were) and groups of observations, as defined by the distinct values of the variable specified in over() or by(), are placed more broadly. What is to be done if the opposite order is wanted? Suppose that the contrast between January and July (Northern Hemisphere winter and summer) is thought less interesting than the contrasts between different regions. We then need regions, not months, to be next to each other.

For the opposite order, we need a different data structure, which can be obtained through the reshape command. If reshape is new to you, refer to the online help and manual entry. In this example, reshape stacks different variables into one variable that is subdivided by a group variable indicating where the groups came from. This is an easy change of data structure to envisage and one that is often needed.

The citytemp data lack an identifier variable naming the observations, here cities. We do need an identifier for reshape, but the observation number will work well.

```
. generate id = _n
```

In a very large dataset, we would make such an identifier of long storage type. Some judicious renaming of variables can also be a good idea:

```
. rename (tempjan tempjul) (tempJanuary tempJuly)
. reshape long temp, i(id) j(month) string
```

Now the combined variable temp can be grouped by region, as before, and also by month, a new variable created by reshape. We can choose which variable goes on the inside. In this example, we already suspect that comparing temperatures by region may be more interesting than comparing by month. With many other datasets (for example, medical results compared by sex and age group), you may need to experiment to see what works best. Comparisons between subtle effects of interest and starker but well-known effects often recur. Figure 4 shows the results of this example.

```
. graph box temp, over(region) by(month, rows(1) compact note(""))
> ytitle(`title') ylabel(14(18)86, angle(h) grid) `incant1'
. graph hbox temp, over(region) by(month, cols(1) compact note(""))
> ytitle(`title') ylabel(14(18)86, grid) `incant2'
```

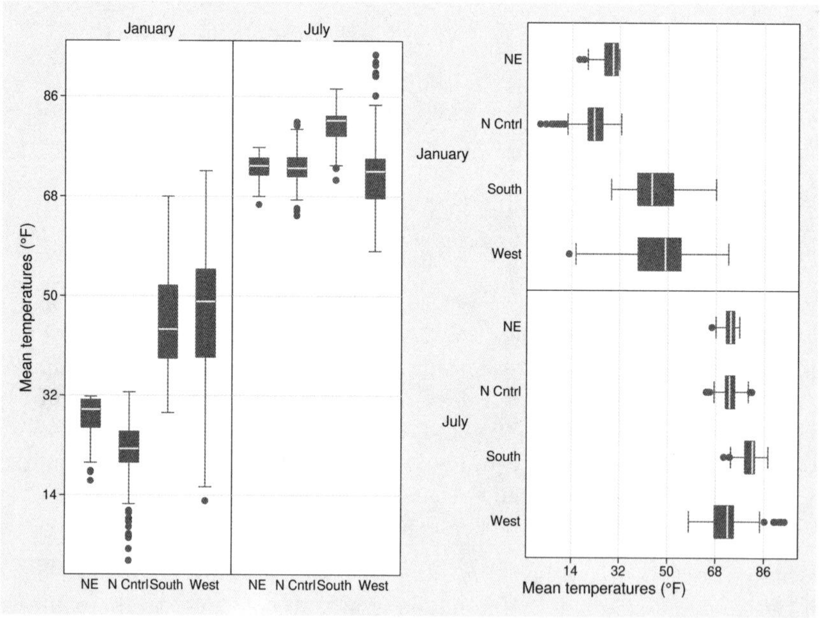

Figure 4. Box plots for January and July temperatures of various U.S. cities using both over() and by() options to compare different regions and months

References

Cox, N. J. 2004. Stata tip 6: Inserting awkward characters in the plot. *Stata Journal* 4: 95–96. https://doi.org/10.1177/1536867X0100400110.

———. 2009. Speaking Stata: Creating and varying box plots. *Stata Journal* 9: 478–496. https://doi.org/10.1177/1536867X0900900309.

———. 2013. Speaking Stata: Creating and varying box plots: Correction. *Stata Journal* 13: 398–400. https://doi.org/10.1177/1536867X1301300214.

10 DOI: 10.1177/1536867X1501500120

The Stata Journal (2015)
15, Number 1, pp. 316–318

Stata tip 122: Variable bar widths in two-way graphs

Ben Jann
University of Bern
Bern, Switzerland
jann@soz.unibe.ch

Two-way bar charts in Stata use a fixed bar width as specified by option `barwidth()` (see [G-2] **graph twoway bar**). Some types of plots found in the literature, however, require variable bar widths. One example is equal probability histograms in which the bar widths are adjusted so that each bar covers the same area; see the `eqprhistogram` command by Cox (1999a). Another example is spine plots for two-way categorical data as implemented in `spineplot` (Cox 2008, 2014).

In this tip, I highlight the `bartype(spanning)` option of the `twoway bar` command, an undocumented feature that can be used to produce bars of different widths. In fact, `eqprhistogram` and `spineplot` are based on this functionality. Consider a plot of income or wealth shares by population percentiles, as is sometimes used in inequality research.[1] Such a plot could be produced as follows:

```
. sysuse nlsw88
(NLSW, 1988 extract)
. sort wage
. generate cumwage = sum(wage)
. replace  cumwage = cumwage/cumwage[_N]
(2246 real changes made)
. _pctile  cumwage, percentiles(20 40 60 70 80 90 95 97 99)
. return list
scalars:
                  r(r1) =  .0803691893815994
                  r(r2) =  .2007369846105576
                  r(r3) =  .3634746670722961
                  r(r4) =  .4661617577075958
                  r(r5) =  .5882071852684021
                  r(r6) =  .7350806593894958
                  r(r7) =  .8268628716468811
                  r(r8) =  .8735467791557312
                  r(r9) =  .9496700167655945
```

1. For an amazing example, see http://www.youtube.com/watch?v=slTF_XXoKAQ. The percentile share plot is a binned and rescaled version of the quantile plot (see [R] **diagnostic plots** and Cox [1999b]). In inequality research, the quantile plot is also known as Pen's "Parade of Dwarfs (and a few Giants)" (Pen 1971, 48–59).

```
. matrix S = ( 0,   (r(r1) - 0     ) / ( 20 - 0 ) * 100)
>            \ ( 20, (r(r2) - r(r1)) / ( 40 - 20) * 100)
>            \ ( 40, (r(r3) - r(r2)) / ( 60 - 40) * 100)
>            \ ( 60, (r(r4) - r(r3)) / ( 70 - 60) * 100)
>            \ ( 70, (r(r5) - r(r4)) / ( 80 - 70) * 100)
>            \ ( 80, (r(r6) - r(r5)) / ( 90 - 80) * 100)
>            \ ( 90, (r(r7) - r(r6)) / ( 95 - 90) * 100)
>            \ ( 95, (r(r8) - r(r7)) / ( 97 - 95) * 100)
>            \ ( 97, (r(r9) - r(r8)) / ( 99 - 97) * 100)
>            \ ( 99, (     1 - r(r9)) / (100 - 99) * 100)
>            \ (100, .)

. matrix list S

S[11,2]
            c1          c2
 r1          0   .40184595
 r2         20   .60183898
 r3         40   .81368841
 r4         60   1.0268709
 r5         70   1.2204543
 r6         80   1.4687347
 r7         90   1.8356442
 r8         95   2.3341954
 r9         97   3.8061619
r10         99   5.0329983
r11        100           .

. svmat S

. twoway bar S2 S1, bartype(spanning)
> yline(1) xtitle(Percentile) ytitle(Income share)
```

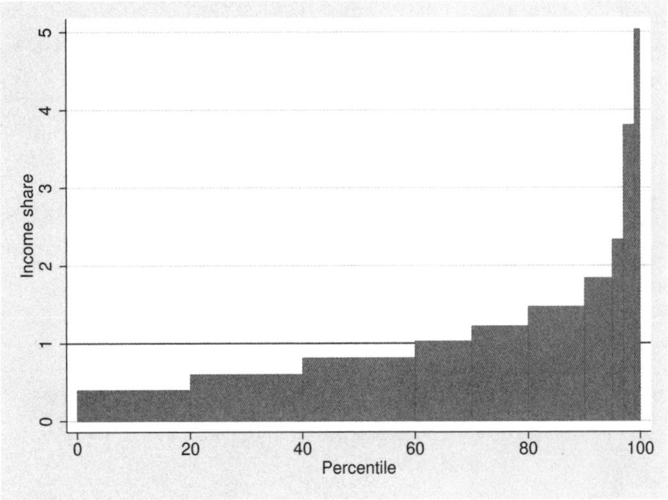

First, the running sum of ordered wages is computed and divided by the wage total. Second, `_pctile` is used to compute a series of percentiles from the cumulated wages (see [D] **pctile**). Results are collected in a matrix, where income shares are computed as differences between consecutive percentiles and normalized by the population share.

Third, `svmat` (see [P] **matrix mkmat**) is used to store the matrix columns as variables, and `twoway bar` is applied with the `bartype(spanning)` option to create the graph. Variable `S1` (the first column in the matrix) specifies the lower bounds of the bars on the x axis, and variable `S2` (the second column) specifies the heights of the bars. The width of a bar is determined by `bartype(spanning)` such that it spans the x axis to the lower bound of the next bar. An extra row in the matrix is needed to provide the upper bound for the last (rightmost) bar.

Such a plot provides a very intuitive view on the wage distribution. Suppose you have 100 dollars to distribute among 100 people. The people are lined up along the x axis in ascending order of their shares, which are depicted by the heights of the bars. If the distribution is equal, each person gets 1 dollar (horizontal line). For the data at hand, however, we see that there is inequality. For example, the person with the largest share gets 5 dollars, and the 20 people with the lowest shares get only about 40 cents on average. The area of a bar reflects the fraction of total wages received by the corresponding group. For example, the bottom 20% of people receive $0.4 \times 20 = 8\%$ of the sum of wages.

References

Cox, N. J. 1999a. eqprhistogram: Stata module for equal probability histogram. Statistical Software Components S432701, Department of Economics, Boston College. https://ideas.repec.org/c/boc/bocode/s432701.html.

———. 1999b. gr42: Quantile plots, generalized. *Stata Technical Bulletin* 51: 16–18. Reprinted in *Stata Technical Bulletin Reprints*. Vol. 9, pp. 113–116. College Station, TX: Stata Press.

———. 2008. Speaking Stata: Spineplots and their kin. *Stata Journal* 8: 105–121. https://doi.org/10.1177/1536867X0800800107.

———. 2014. *Speaking Stata Graphics*. College Station, TX: Stata Press.

Pen, J. 1971. *Income Distribution*. London: Allen Lane.

The Stata Journal (2015)
15, Number 1, pp. 319–323 DOI: 10.1177/1536867X1501500121

Stata tip 123: Spell boundaries

Nicholas J. Cox
Department of Geography
Durham University
Durham, UK
n.j.cox@durham.ac.uk

1 Introduction

Identifying spells or runs of observations is a common problem in data management and data summary. A detailed tutorial was given in a "Speaking Stata" column (Cox 2007). That column identified some simple techniques for working with spells in Stata:

1. Mark the start of each spell with an indicator variable. The key is that observations at the start of spells will differ from their predecessors. Care may be needed in handling the very first observation, either in a dataset or in a panel.
2. Use cumulative sums to map start indicators to spell identifiers that are 1 up. It is also useful to identify gaps between spells by 0. Given identifiers, summarizing spell characteristics is then usually straightforward. `egen` functions are particularly useful.
3. Panel datasets are no more difficult than individual series, so long as you use `by:`. Using features allowed after `tsset` or `xtset` is perfectly sensible but not essential.
4. Some spell criteria do require two passes through the data. Typically, spells are reclassified on the second pass, say, to restrict spells to certain lengths or to allow short gaps within spells.

If you are nodding in agreement with these points, do read on. If they seem cryptic, please read (or skim) the 2007 column first.

Here I add further detail on two common problems. The first is when a spell is defined by its end condition. With just a twist, this can be recast easily as a condition for defining its start.

The second problem need not be considered as a question of defining spells but can be seen in that framework. This problem pertains to calculating the time or number of observations since some event, which can be approached directly.

2 The ends define the spells

Many spells are defined just as naturally by when they end as by when they start. Sometimes, the exact time of ending may be known, but the starting time may be unclear or definable only arbitrarily. For example, an election marks the end of a political campaign. A birth marks the end of a pregnancy. A sale may mark the end of a period of contact between a seller and a potential buyer.

Let's assume that we have, or can create, an indicator variable for the end of a spell, say, `end`. Let's also assume that data are sorted in order of a time or other sequencing variable, possibly within panels in the case of panel or longitudinal data. The criterion for the start of a spell can then be something like

```
. generate begin = end[_n-1] | _n == 1
```

for a single series. For panel data, it may be

```
. bysort id (time): generate begin = end[_n-1] | _n == 1
```

Here the code is short for

```
. generate begin = (end[_n-1] == 1) | (_n == 1)
```

and similarly for the panel case. If the indicator variable is only ever 1 or 0, then `end[_n-1] == 1` yields 1 (true) precisely when `end[_n-1]` is 1 (true).

In general, two possibilities define the start of a spell. Either the previous observation was the end of a spell, or this is the first observation. Here `_n` defines observation number; under the aegis of `by:`, observation numbers are defined within groups; and `|` is the logical operator "or".

Given that indicator variable for the beginning of a spell, a spell identifier is just

```
. generate spellid = sum(begin)
```

Let's see how this works with a small example. There are 7 observations, and the indicator `end` is 1 in observation 4.

```
. list, sep(0)
```

	time	end	begin	spellid
1.	1	0	1	1
2.	2	0	0	1
3.	3	0	0	1
4.	4	1	0	1
5.	5	0	1	2
6.	6	0	0	2
7.	7	0	0	2

Some complications of this basic idea are predictable. With this approach, a spell will always be identified with identifier 1, regardless of whether the event took place. For example, the potential buyer may never have proceeded to purchase. A spell was identified as such because it started with the first observation. In that case, it may be sensible to reclassify the spell because it was incomplete. That is easy. For a single series, we can summarize end and replace spellid if its maximum is only 0. After summarize, the maximum is accessible as r(max).

```
. summarize end, meanonly
. if r(max) == 0 replace spellid = 0
```

Here we reclassified an incomplete spell to have 0 as an identifier. Another possibility that might appeal is to reclassify to a missing value.

For panel data, we need to examine each panel separately. We can calculate the maximum of end for each panel with

```
. bysort id: egen max = max(end)
. replace spellid = 0 if max == 0
```

and thus reclassify the incomplete spell in each panel.

A similar problem arises if the last spell—whether for all the data or for a single panel—is incomplete. Again, "incomplete" means that the last observation has a value of 0 for end. For a single series, we reclassify with

```
. replace spellid = 0 if end[_N] == 0 & spellid == spellid[_N]
```

and for panel data, we merely apply that under by:,

```
. bysort id (time): replace spellid = 0 if end[_N] == 0 & spellid == spellid[_N]
```

However, notice now that this technique will take care of the first problem too. If the data define a single incomplete spell, then it will also be true that the last observation has value 0 for the end indicator, and the condition that the spell identifier equals the last spell identifier will catch all the relevant data. So you can forget the first technique, or feel good about having two ways of solving the problem.

3 Time since an event

Researchers often want to keep track of the time since some event. Events can be anything deemed to happen at a single time or point in a sequence. A common example on Statalist is an initial public offering or a stock market launch.

In a sense, this problem is a twist on the problem of identifying the previous value of some variable that was different (Cox 2011), but the "Tip" just cited did not spell that out, and the problem fits in here quite well.

It is also a twist on the problem of identifying a counter variable indicating sequence within spells, which is a problem also discussed in Cox (2007). But we can approach it without even identifying spells explicitly, and will do so now.

We again assume an indicator variable for an event and order by time or sequence variable, possibly within panels, as in the previous section. We need not be precise about whether the event marks the start or end of a spell. Hence, we phrase matters using a neutral variable name such as `event`. `event` will be 1 when an event occurred and 0 otherwise. The times at which events occur can be copied into a new variable with

```
. generate when = time if event
```

which is a valid command for either a single series or panel data. Note that the condition `if event` implies that values will be missing for times other than those when an event took place. But we can copy values downward in a cascade. For a single series, we use

```
. replace when = when[_n-1] if missing(when)
```

and for panel data, we use

```
. bysort id (time): replace when = when[_n-1] if missing(when)
```

If you are not familiar with this trick, it works like this: to begin, `generate` and `replace` use the current order of observations (Newson 2004). So a missing value immediately after a nonmissing value of `when` can be `replaced` with that nonmissing value; a missing value immediately after that can be `replaced` with the same nonmissing value, which is now the previous value. We then continue in a cascade until we reach the next nonmissing value or the end of the panel or the end of the data, whichever comes first.

Now the times since the last event are immediately available by subtraction, as follows:

```
. generate time_since = time - when
```

Sometimes, researchers like to restrict attention or calculation to times within some specified interval of an event, say, within the next 30 days or 2 years. The variable just created will then naturally appear in a condition specified with `if`, such as `time_since <= 30`.

An inevitable side effect of this calculation is that the variables `when` and `time_since` will be returned with missing values for observations before the first event. Typically, that should be considered logical and desirable.

An attraction of this device is that few assumptions are being made. There is no assumption that times are evenly spaced. There is no assumption that a time variable or a panel identifier variable has been declared with `tsset` or `xtset`.

A variant on the problem is that researchers sometimes want to count observations after an event rather than measure the time elapsed. The two are not equivalent whenever times are not evenly spaced or the difference between times is not 1 in whatever time units are being used. An example might be counting patient visits to a clinic after some event, say, an initial consultation.

A technique for this preference is to initialize a counter, as follows:

```
. generate counter = 0 if event
```

We then count upward (as in elementary arithmetic) by adding 1 repeatedly. For single series, use

```
. replace counter = counter[_n-1] + 1 if missing(counter)
```

For panel data, do that within panels. If it makes more sense to regard the event itself as a count of 1, the modification is clear.

References

Cox, N. J. 2007. Speaking Stata: Identifying spells. *Stata Journal* 7: 249–265. https://doi.org/10.1177/1536867X0700700209.

———. 2011. Stata tip 101: Previous but different. *Stata Journal* 11: 472–473. https://doi.org/10.1177/1536867X1101100309.

Newson, R. B. 2004. Stata tip 13: generate and replace use the current sort order. *Stata Journal* 4: 484–485. https://doi.org/10.1177/1536867X0400400411.

 DOI: 10.1177/1536867X1501500218

The Stata Journal (2015)
15, Number 2, pp. 597–598

Stata tip 124: Passing temporary variables to subprograms

Maarten L. Buis
Department of History and Sociology
University of Konstanz
Konstanz, Germany
maarten.buis@uni-konstanz.de

A useful tool when programming in Stata is the temporary variable, which can be created using the `tempvar` command (see [P] **macro**). When it is convenient to store intermediate steps in a temporary variable, `tempvar` reserves a variable name for that temporary variable that is guaranteed not to exist in your current dataset. This ensures that your program will not accidentally overwrite an already existing variable. `tempvar` also ensures that the temporary variable is removed once the program that created it is finished so that your program will not clutter the user's dataset with unwanted intermediate results. Similarly, one can create temporary scalars and matrices with the `tempname` command (see [P] **macro**). When one programs in Stata, it is also useful to break up larger programs into various smaller subroutines. This helps to keep longer programs organized and makes it easier to write, debug, certify, and maintain them. Sometimes, creating temporary results in a temporary variable is a good candidate for such a subroutine. If we use `tempvar` or `tempname` in that subroutine, the temporary variable, scalar, or matrix will be deleted as soon as the subroutine is finished. In this case, that is not what we want.

To use the temporary objects created or changed in subroutines in the main program, we need to use `tempvar` or `tempname` in the main program and pass that name to the subroutine. Consider the example below.

```
. set seed 1234567

. program mainprog
  1.         tempvar random
  2.         quietly generate `random' = .
  3.         tempname mean
  4.         scalar `mean' = 2
  5.         subprog, random(`random') mean(`mean')
  6.         summarize `random'
  7. end

. program subprog
  1.         syntax, random(name) mean(name)
  2.         quietly replace `random' = rnormal(`mean')
  3. end

. sysuse auto
(1978 Automobile Data)

. mainprog

    Variable |        Obs        Mean    Std. Dev.       Min        Max
-------------+---------------------------------------------------------
    __000000 |         74     1.87675    1.013603  -.4361137   4.350792
```

In line 1 of `mainprog`, a variable name is chosen that does not exist in the current data, and this variable name is stored in the local macro `` `random' ``. In line 2, this name is used to create a variable. In lines 3 and 4, a temporary scalar `` `mean' `` is created. In line 5, the names of the temporary variable and the temporary scalar are passed to `subprog` in the options `random()` and `mean()`. Notice that `subprog` runs when `mainprog` is not yet finished, so variables created with `tempvar` and matrices and scalars created with `tempname` still exist. Line 1 of `subprog` means that `subprog` expects two options containing a name, and that name will be stored in the local macros `` `random' `` and `` `mean' ``. Line 2 of `subprog` then changes the temporary variable by using the temporary scalar. Now, we go back to line 6 of `mainprog`, which uses that changed temporary variable. `mainprog` ends, and the temporary variable `` `random' `` and temporary scalar `` `mean' `` are deleted.

The same logic can also be used to pass temporary variables, matrices, and scalars to Mata functions; as long as the program that created them has not finished, the objects exist. To pass them on, you must pass their names to the Mata function. For example, the program below does the same thing as the example above, except that it uses Mata for the subroutine.

```
. clear all

. mata
------------------------------------------------- mata (type end to exit) -----
: void mata_subprog(
>     string scalar randomname,
>     string scalar meanname) {
>
>     st_view(random=., ., randomname)
>     mean = st_numscalar(meanname)
>
>     random[.,.] = rnormal(st_nobs(),1,mean,1)
> }

: end
-------------------------------------------------------------------------------

. program mainprog
  1.          tempvar random
  2.          quietly generate `random' = .
  3.          tempname mean
  4.          scalar `mean' = 2
  5.          mata: mata_subprog("`random'", "`mean'")
  6.          summarize `random'
  7. end

. sysuse auto
(1978 Automobile Data)

. mainprog

    Variable |        Obs        Mean    Std. Dev.       Min        Max
-------------+---------------------------------------------------------
     __000000 |         74    1.904863    .9529203  -.5836316   3.828233
```

20 DOI: 10.1177/1536867X1501500219

The Stata Journal (2015)
15, Number 2, pp. 599–604

Stata tip 125: Binned residual plots for assessing the fit of regression models for binary outcomes

Jessica Kasza
Monash University
Melbourne, Australia
jessica.kasza@monash.edu

Plots based on residuals, such as plots of residual-versus-fitted values, are now standard after fitting linear regression models. These plots are used to assess the validity of assumptions, to identify features not captured by the model, and to find problematic data points or clusters. However, such plots are typically not very useful for regression models for binary outcomes because of the discrete nature of residuals from these models. Binned residual plots, as recommended by Gelman and Hill (2007), can be used to assess both the overall fit of regression models for binary outcomes (for example, logistic or probit models) and the inclusion of continuous variables. I demonstrate the construction of such plots in Stata. These binned residual plots are related to those produced by the `rbinplot` command from the `modeldiag` package (Cox [2004], updated in Cox [2010]), with the addition of approximate confidence limits.

To construct a binned residual plot to assess the overall fit of a logistic regression model, one orders predicted probabilities from smallest to largest and calculates residuals. Data are split into bins containing equal numbers of observations (a recommended number of bins is the square root of the number of observations), and the average residual is plotted against the average predicted probability for each bin. For each bin, approximate 95% confidence limits are $\pm 2\sqrt{p(1-p)/n}$, estimated using the standard deviation of each bin's residuals.

If the model is correct, about 95% of the points are expected to lie within the confidence limits. As is the case for a residual-versus-fitted plot used for linear regression, departures from random scatter are indicative that the fitted model does not accurately describe the data. To assess the fit of a continuous covariate, one orders observations and constructs bins in terms of that covariate instead of in terms of the predicted probabilities. The average residual is then plotted against the average covariate in each bin.

To demonstrate the construction of these plots, I simulate an example dataset consisting of 5,000 observations, where the binary outcome (or response variable) is dependent on two continuous covariates and the square of one of these covariates:

```
. set obs 5000
obs was 0, now 5000
. set seed 86206
. generate x1 = rnormal()
. generate x2 = rnormal()
. generate prob_y = exp(-1+x1+x2+x1^2)/(1+exp(-1+x1+x2+x1^2))
. generate y = rbinomial(1, prob_y)
```

To demonstrate the usefulness of binned residual plots, I omit the squared term from the logistic regression model for the binary outcome.

```
. quietly logit y x1 x2
. predict pred_y, pr
. generate resid = y - pred_y
```

After I obtain the predicted probabilities for each observation, the construction of the binned residual plot proceeds as follows:

```
. sort pred_y
. generate myids = _n if pred_y < .
. local nbins = floor(sqrt(5000))
. egen binno = cut(myids) if pred_y < . , group(`nbins´) icodes
. egen avefit = mean(pred_y), by(binno)
. egen myaveres = mean(resid), by(binno)
. egen mysd = sd(resid), by(binno)
. egen mytag = tag(binno)
. bysort binno: egen binsize = count(pred_y)
. generate uplim = 2*mysd/sqrt(binsize)
. generate dwlim = -2*mysd/sqrt(binsize)
. graph twoway (scatter myaveres avefit if inrange(myaveres, dwlim, uplim)
> & mytag == 1, msymbol(oh))
> (line uplim avefit, clcolor(gray) lstyle(solid))
> (line  dwlim avefit, clcolor(gray) lstyle(solid))
> (scatter myaveres avefit if !inrange(myaveres, dwlim, uplim) & mytag == 1,
> mcolor(black)),
> xtitle(Average predicted mortality probability) ytitle(Average residual)
> legend(off) scheme(sj)
```

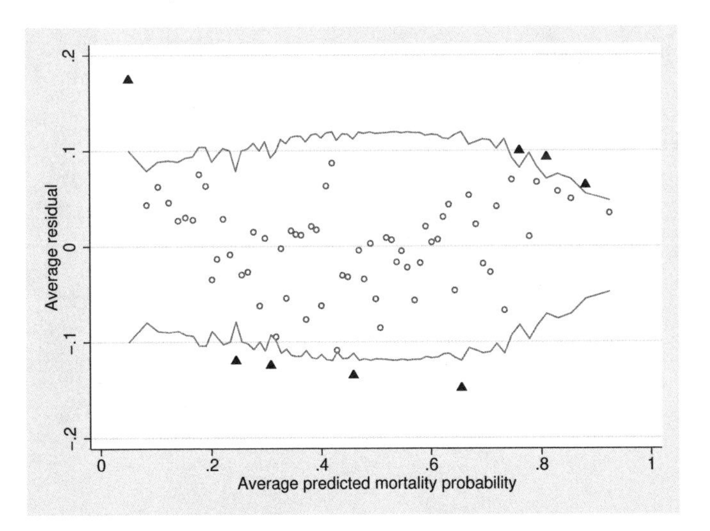

Figure 1. Binned residual plot to assess the overall fit of the model

Figure 1 displays the produced binned residual plot. There is some curvature to the pattern of binned residuals, although this is not particularly extreme. However, assessing the fit of the model with respect to `x1` does indicate problems.

```
. keep y x1 x2 pred_y resid
. sort x1
. generate myids = _n if pred_y < .
. local nbins = floor(sqrt(5000))
. egen binno = cut(myids) if pred_y < . , group(`nbins´) icodes
. egen avex1 = mean(x1), by(binno)
. egen myaveres = mean(resid), by(binno)
. egen mysd = sd(resid), by(binno)
. egen mytag = tag(binno)
. bysort binno: egen binsize = count(pred_y)
. generate uplim = 2*mysd/sqrt(binsize)
. generate dwlim = -2*mysd/sqrt(binsize)
. graph twoway (scatter myaveres avex1  if inrange(myaveres, dwlim, uplim) &
> mytag == 1, msymbol(oh))
> (line uplim avex1, clcolor(gray) lstyle(solid))
> (line dwlim avex1, clcolor(gray) lstyle(solid))
> (scatter myaveres avex1 if !inrange(myaveres, dwlim, uplim) & mytag == 1,
> mcolor(black)),
> xtitle(Average x1) ytitle(Average residual) legend(off) scheme(sj)
```

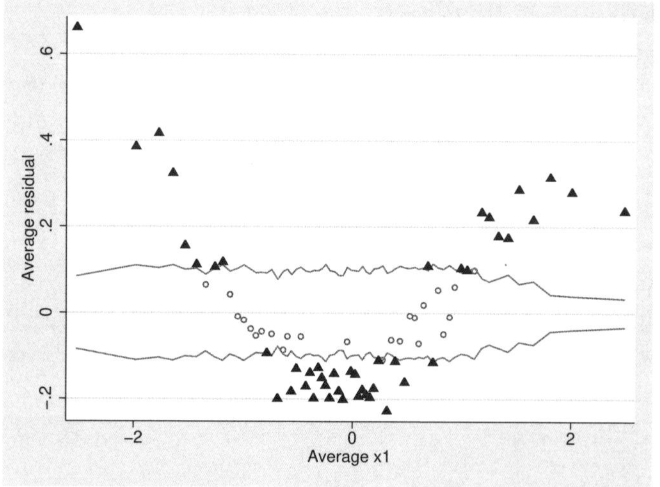

Figure 2. Binned residual plot to assess the fit of `x1`

The systematic pattern in figure 2 indicates that nonlinear terms for `x1` should be included in the logistic regression model. We fit the model including a quadratic term for `x1` as follows:

```
. keep y x1 x2
. qui logit y x1 x2 c.x1#c.x1 c.x2#c.x2
. predict pred_y
(option pr assumed; Pr(y))
. generate resid = y - pred_y
```

For this model, binned residual plots are constructed as above and displayed in figure 3. As is expected, a few points lie outside the confidence limits, but there are no systematic patterns in the plots.

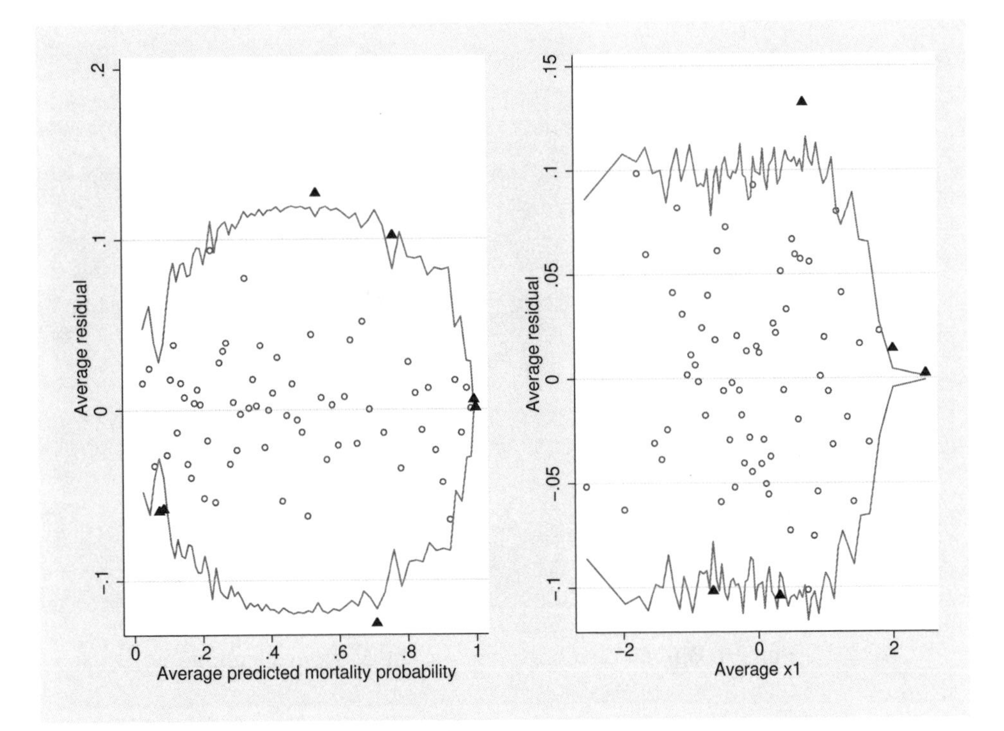

Figure 3. Binned residual plots for the model containing a quadratic term for `x1`

To illustrate the usefulness of binned residual plots, we consider the Medpar dataset from Hilbe (2009), available at http://www.crcpress.com/product/isbn/9781420075755. This dataset is a subset of the 1991 U.S. national Medicare inpatient hospital database for Arizona, and it consists of data from 1,495 randomly selected patients. The first model for in-hospital mortality included length of hospital stay, indicators for age over 80 years, and type of surgery (elective, urgent, or emergency, with elective as baseline):

```
. use medpar.dta, clear
. quietly logit died los age80 type2 type3
. predict pred_y, pr
. generate resid = died - pred_y
. sort pred_y, stable
```

Binned residual plots to assess the overall fit and the inclusion of length of stay were constructed and are displayed in the first row of figure 4. Because many patients have identical estimated mortality probabilities, the `stable` option of the `sort` command should be used.

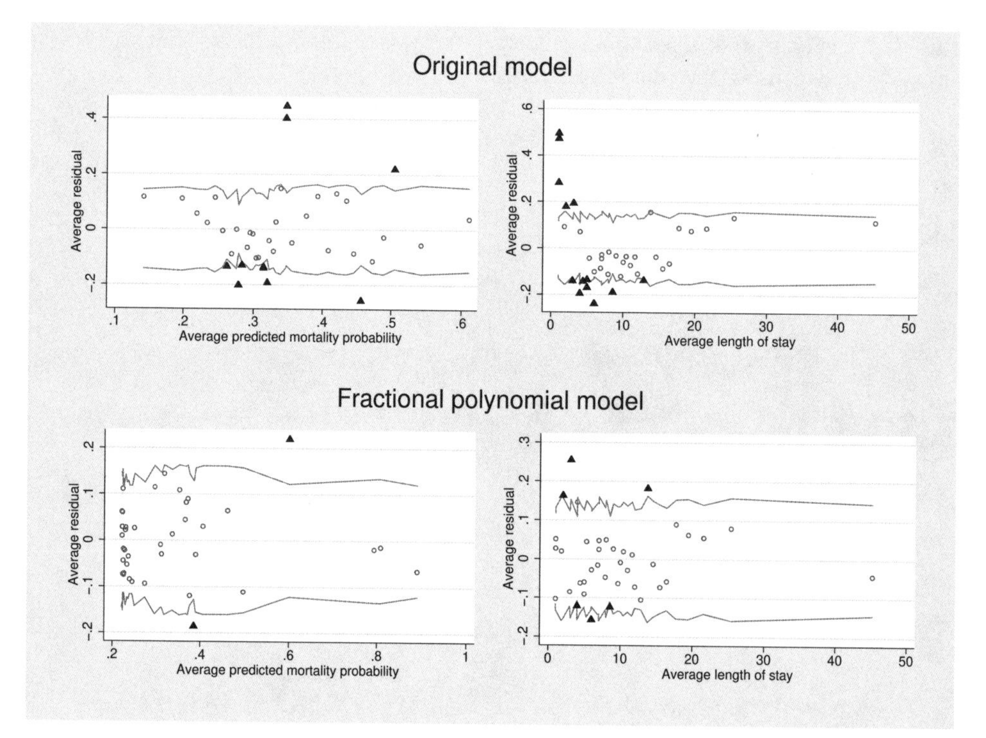

Figure 4. Binned residual plots for the Medpar example

The binned residual plot for length of stay indicates that a linear term for length of stay does not accurately capture the relationship between mortality and length of stay. A multivariable fractional polynomial logistic model is fit as follows:

```
. quietly mfp logit died los age80 type2 type3
```

The selected model contains indicators for age over 80 years and type of surgery, and a degree-1 fractional polynomial for length of stay with power -2. Binned residual plots to assess the overall fit and length of stay are displayed in the second row of figure 4. These plots indicate that this model is a much better fit to the data than that containing a linear term for length of stay.

References

Cox, N. J. 2004. Speaking Stata: Graphing model diagnostics. *Stata Journal* 4: 449–475. https://doi.org/10.1177/1536867X0400400408.

———. 2010. Software Updates: gr0009_1: Speaking Stata: Graphing model diagnostics. *Stata Journal* 10: 164. https://doi.org/10.1177/1536867X1001000116.

Gelman, A., and J. Hill. 2007. *Data Analysis Using Regression and Multilevel/Hierarchical Models.* Cambridge: Cambridge University Press. https://doi.org/10.1017/CBO9780511790942.

Hilbe, J. M. 2009. *Logistic Regression Models.* Boca Raton, FL: Chapman and Hall/CRC. https://doi.org/10.1201/9781420075779.

26 DOI: 10.1177/1536867X1601600216

The Stata Journal (2016)
16, Number 2, pp. 517–520

Stata tip 126: Handling irregularly spaced high-frequency transactions data

Christopher F. Baum
Department of Economics
Boston College
Chestnut Hill, MA
baum@bc.edu

Sebastiaan Bibo
University of Amsterdam
Amsterdam, Netherlands

A wealth of high-frequency data is available on the Internet for many financial markets. Many of these datasets are "tick data", or transactions data, for which one observation represents one trade. Trades may occur at irregular intervals in calendar time, depending on market activity. To study these data—in particular, to analyze data from more than one market produced in the same time frame—one must convert irregularly spaced transactions data to time intervals. The observation for one time interval usually contains the number of trades during the interval (which could be zero), the average price of the good traded, and both the high and the low price during the interval. Other measures, such as standard deviations for each interval or total trade volume, may also be recorded.

In this tip, we illustrate how a transactions-based financial dataset of bitcoin trades can be converted to a dataset of standardized time intervals so that these datasets from multiple markets may be juxtaposed to evaluate the presence of arbitrage opportunities, which occur when the same good sells for different prices in different markets. That is only a necessary condition: if arbitrage is to be profitable, transactions costs and bid-ask spreads must be considered.

We begin by downloading the raw data for one market from http://api.bitcoincharts.com/v1/csv/bitstampUSD.csv.gz and by using the `gzip` program[1] to extract the `.csv` formatted data.[2]

```
. !gzip -vd bitstampUSD.csv.gz
bitstampUSD.csv.gz:      83.5% -- replaced with bitstampUSD.csv
```

We can now import the data, using the `asdouble` option to ensure that the integer timestamp is preserved to its full precision. The first column of the input data is the date and time of the transaction, recorded as a Unix timestamp. Unix timestamps count the number of seconds since midnight, 1 January 1970. To create a Stata clocktime variable with format `%tc`, we must scale the Unix timestamp by 1,000 to express it in milliseconds, and we must set the base date for the Stata variable as 1/1/1970. For this market, we have about 8.67 million transactions recorded as of mid-January 2016.

1. `gzip` is built-in on Linux, Unix, and Mac OS X systems.
2. Stata's `unzipfile` command does not deal with `gzip` files. If your system lacks `gzip`, see http://www.gzip.org.

```
. import delimited bitstampUSD.csv, asdouble
(3 vars, 8,671,372 obs)
. generate double tstamp = (v1 * 1000) + mdyhms(1,1,1970,0,0,0)
. format tstamp %tc
. summarize tstamp, f

    Variable |        Obs        Mean    Std. Dev.       Min        Max
-------------+---------------------------------------------------------
      tstamp |  8,671,372  03aug2014 18:36:19   2.46e+10  13sep2011 13:53:36
> 19jan2016 11:45:20
. rename (v2 v3) (price volume)
. save "bitstampUSD.dta", replace
file bitstampUSD.dta saved
```

We now must specify the range of dates to be studied and the time interval in which transactions are to be collapsed. To illustrate, we choose 1 January 2014–31 December 2015 as the time period, to be expressed in five-minute intervals. We create the clocktime values for the beginning and ending dates as scalars,[3] and we compute the number of new observations needed to provide the regular intervals. Here we have 210,240 five-minute intervals in these two calendar years.

```
. describe
Contains data from bitstampUSD.dta
  obs:     8,671,372
 vars:             4                          6 Mar 2016 16:00
 size:   242,798,416
------------------------------------------------------------------------------
              storage   display    value
variable name   type    format     label      variable label
------------------------------------------------------------------------------
v1              long    %12.0g
price           double  %10.0g
volume          double  %10.0g
tstamp          double  %tc
------------------------------------------------------------------------------
Sorted by:
. scalar tobs = r(N)
. scalar firstdate = mdyhms(1,1,2014,0,0,0)
. scalar lastdate  = mdyhms(12,31,2015,23,59,59)
. scalar interval  = 5 * 60 * 1000  // ms in 5 minute interval
. scalar addobs = (lastdate - firstdate) / interval
. display _n "interval observations = " addobs
interval observations = 210240
. scalar newobs = int(tobs + addobs)
```

We now expand the dataset by setting the new number of observations, and we fill in the `tstamp` values for the new observations. The transactions outside the range are dropped. We then sort the data by `tstamp` to intersperse the interval observations among the transactions data.

3. If you are unfamiliar with Stata's scalars, see [P] **scalar**.

```
. set obs `=newobs´
number of observations (_N) was 8,671,372, now 8,881,611
. local t1 = tobs + 1
. local t2 = tobs + 2
. replace tstamp = firstdate + interval in `t1´
(1 real change made)
. replace tstamp = tstamp[_n-1] + interval in `t2´/L
(210,238 real changes made)
. drop if tstamp < firstdate
(2,222,393 observations deleted)
. drop if tstamp > lastdate
(123,400 observations deleted)
. sort tstamp
```

Producing the regularly spaced interval data is now straightforward. We create an indicator variable, `intvl`, equal to one in each interval observation. The `sum()` function produces a running sum in `wintvl`, numbering each interval sequentially. This variable is then used to drive the `collapse` command, which produces a new dataset with one observation per interval. Each observation is identified by the starting date and time of the interval. We can also now `tsset` the data, specifying a `delta` of five minutes. If there are concerns over missing data for intervals in which no transactions occurred, [D] **ipolate** or another interpolation method could be used to fill in the price series.

```
. generate wintvl = sum(missing(price)) + 1
. collapse (min) tstamp price (sum) volume (min) minp = price (max) maxp = price
> (sd) sdprice = price (count) ntrans = price, by(wintvl)
. tsset tstamp, delta(5 minutes)
        time variable:  tstamp, 01jan2014 00:00:00 to 31dec2015 23:55:00
                delta:  5 minutes
. list in 1/10, sep(0) noobs
```

wintvl	tstamp	price	volume	minp	maxp
1	01jan2014 00:00:00	729.01	57.979092	729.01	734

sdprice	ntrans
1.3688354	38

wintvl	tstamp	price	volume	minp	maxp
2	01jan2014 00:05:00	730.33	19.98457	730.33	734.34

sdprice	ntrans
1.5530906	23

wintvl	tstamp	price	volume	minp	maxp
3	01jan2014 00:10:00	733.7	20.442605	733.7	734.5

sdprice	ntrans
.35482983	20

```
  +------------------------------------------------------------------------+
  | wintvl |              tstamp |  price |    volume |   minp |     maxp |
  |      4 | 01jan2014 00:15:00  | 732.19 |  34.47609 | 732.19 |      735 |
  |------------------------------------------------------------------------|
  |                 sdprice                  |             ntrans          |
  |               1.0303945                  |                 31          |
  +------------------------------------------------------------------------+

  +------------------------------------------------------------------------+
  | wintvl |              tstamp |  price |    volume |   minp |     maxp |
  |      5 | 01jan2014 00:20:00  | 731.81 | 127.53413 | 731.81 |      738 |
  |------------------------------------------------------------------------|
  |                 sdprice                  |             ntrans          |
  |               2.1149646                  |                 83          |
  +------------------------------------------------------------------------+

  +------------------------------------------------------------------------+
  | wintvl |              tstamp |  price |    volume |   minp |     maxp |
  |      6 | 01jan2014 00:25:00  | 734.56 |   5.76307 | 734.56 |   738.25 |
  |------------------------------------------------------------------------|
  |                 sdprice                  |             ntrans          |
  |               1.5949494                  |                 12          |
  +------------------------------------------------------------------------+

  +------------------------------------------------------------------------+
  | wintvl |              tstamp |  price |    volume |   minp |     maxp |
  |      7 | 01jan2014 00:30:00  | 734.81 | 6.3947002 | 734.81 |   738.23 |
  |------------------------------------------------------------------------|
  |                 sdprice                  |             ntrans          |
  |               1.0703673                  |                  9          |
  +------------------------------------------------------------------------+

  +------------------------------------------------------------------------+
  | wintvl |              tstamp |  price |    volume |   minp |     maxp |
  |      8 | 01jan2014 00:35:00  | 734.81 |  17.86617 | 734.81 |   738.23 |
  |------------------------------------------------------------------------|
  |                 sdprice                  |             ntrans          |
  |               1.3222282                  |                 22          |
  +------------------------------------------------------------------------+

  +------------------------------------------------------------------------+
  | wintvl |              tstamp |  price |    volume |   minp |     maxp |
  |      9 | 01jan2014 00:40:00  | 735.47 | 51.953554 | 735.47 |   738.28 |
  |------------------------------------------------------------------------|
  |                 sdprice                  |             ntrans          |
  |               1.2391635                  |                 23          |
  +------------------------------------------------------------------------+

  +------------------------------------------------------------------------+
  | wintvl |              tstamp |  price |    volume |   minp |     maxp |
  |     10 | 01jan2014 00:45:00  | 734.47 | 195.32778 | 734.47 |      739 |
  |------------------------------------------------------------------------|
  |                 sdprice                  |             ntrans          |
  |               1.6481093                  |                 52          |
  +------------------------------------------------------------------------+

. save bitstampUSD_interval, replace
file bitstampUSD_interval.dta saved
```

To aggregate transactions to intervals of a different length, we need change only the definition of `interval`. Likewise, different ranges of the transactions history may be specified by redefining `firstdate` and `lastdate`.

DOI: 10.1177/1536867X1701700215

The Stata Journal (2017)
17, Number 2, pp. 511–514

Stata tip 127: Use capture noisily groups

Roger B. Newson
Department of Primary Care and Public Health
Imperial College London
London, UK
r.newson@imperial.ac.uk

1 The main idea

Do you know about the `capture noisily` group? It is a group of Stata commands, starting with `capture noisily {` and ending with `}`. The commands in between are executed as usual, producing the standard output (because of `noisily`). If any of these commands fail, then execution resumes with the command immediately following the group (because of `capture`). If you do not like typing `capture noisily`, then you can abbreviate it to `cap noi`, or even to `cap n`.

A simple application is where the user wishes to generate a Stata log file, which the user inspects afterward to see whether the logged commands work (and especially if they do not). In a do-file, the user may open the log file and begin the `capture noisily` block, using the commands

```
log using mylog.log, replace
capture noisily {
```

and then add a sequence of Stata commands, such as

```
sysuse auto, clear
regress mpg weight
predict mpghat
twoway scatter mpg weight || line mpghat weight, sort
```

and then end the `capture noisily` block and close the log file, using the commands

```
}
log close
```

The commands inside the block will then be executed until one of them fails (or until all of them end execution, if none fail), and their output will be stored in the file `mylog.log`. Whether or not any of the intervening commands fail, the log file `mylog.log` will be closed by the `log close` command. The user may then inspect the log file with a text editor and view the results if execution was successful or find out what went wrong otherwise. Typically, the number of Stata commands inside the block will be more than the four used here, and there may be program loops and other complicated programming constructions (see [P] **forvalues** and [P] **foreach**), increasing the probability of a failure somewhere.

2 Application in estimation command files

The capture noisily prefix is commonly used in do-files containing sequences of estimation commands. If the user is worried that one or more of them might fail (possibly because of insufficient observations), then the user may add a capture noisily prefix to each estimation command so that if one estimation command fails, then Stata will resume execution, starting with the next estimation command. If each estimation command is followed by one or more postestimation commands (such as predict or margins), then each estimation command, and its own subsequent postestimation commands, may be placed in its own capture noisily group. That way, if either the estimation command or the postestimation commands fail, then Stata will continue to the next estimation command.

For instance, in auto.dta, a user might want to fit a regression model of mileage (mpg) with respect to each of the car-size variables weight, length, and displacement, together with the factor foreign, indicating whether a car model is made by a non-U.S. company. After each regression model, the user might want to estimate the mean mileages expected if all cars were U.S. models and if all cars were non-U.S. models, assuming that the car-size variable was distributed as in the real-world sample. The code to do this might be as follows:

```
sysuse auto, clear
describe, full
foreach X of var weight length displacement {
  capture noisily {
    regress mpg ibn.foreign `X´, noconst vce(robust)
    margins i.foreign
  }
}
```

As it happens, this code executes without any failed commands (not shown). However, if (for any reason) the analysis with respect to weight had failed, either in the regress command or in the margins command, then Stata would have proceeded to the analysis with respect to length. If the analysis with respect to length had failed, then Stata would have proceeded to the analysis with respect to displacement. This feature of capture noisily blocks can be very useful if the user is executing a long list of multistep analyses, especially if these analyses involve commands with a higher failure probability than regress. Note that if a multistep analysis fails at an earlier command in a capture noisily block, then the later commands in the same capture noisily block are not attempted. Note also that if the multiple analyses are simply the same command executed on multiple by-groups, then the user does not need the capture noisily block, because the user can use statsby (see [D] **statsby**).

3 Application in file-generation programs

`capture noisily` may also be used to good effect in file-generation programs. For example, it is used internally by the `dolog`, `dologx`, and `dotex` packages, which can be downloaded from the Statistical Software Components (SSC) archive, and used to execute a do-file while automatically generating a log file. However, more advanced users may want to write output to an arbitrary file, using the `file` command documented in [P] **file**. For instance, the new file may be a TEX, an HTML, an Extensible Markup Language, or a Rich Text Format file, produced as an automatically generated report for a reproducible-research project. The user may be using a sequence of commands to generate this new file and may want to close the file after executing those commands, whether or not they all work. The generated file will then be available for the user to inspect (although it may be incomplete), and the user will not have to close it manually. The `capture` block may begin with the commands

```
tempname buff1
file open `buff1´ using "myoutput.txt", write text replace
capture noisily {
```

and contain any amount of intervening code, including `file write` statements, such as

```
file write `buff1´ "Hello, world!!!!"
```

and end with the commands

```
}
file close `buff1´
```

In this case, a new file `myoutput.txt` is created with a buffer, whose name is stored in the local macro `buff1`, and filled with output from the intervening `file write` statements. However, if any statement in the intervening code fails, then Stata executes the `file close` statement, and the new file (usually incomplete) is available for the user to inspect.

Alternatively, the `capture noisily` block may be preceded and followed by file opening and closing commands other than `file open` and `file close`. For instance, if the user is generating an HTML file, then the file opening and closing commands might be the `htopen` and `htclose` commands of the `ht` package (see Quintó et al. [2012]) or the `htmlopen` and `htmlclose` commands of the SSC package `htmlutil` (see Newson [2015]). Or if the user is generating a Rich Text Format file, then they might be the `rtfopen` and `rtfclose` commands of the SSC package `rtfutil` (see Newson [2012]). And more user-written file-generating packages are likely to be written on similar lines in the future, possibly for generating files in Extensible Markup Language-based document formats yet to be invented. Such future packages are likely to contain their own file-opening and file-closing commands, suitable for use before and after a `capture noisily` block.

References

Newson, R. 2015. htmlutil: Stata module to provide utilities for writing hypertext markup language (HTML) files. Statistical Software Components S458085, Department of Economics, Boston College. https://ideas.repec.org/c/boc/bocode/s458085.html.

Newson, R. B. 2012. From resultssets to resultstables in Stata. *Stata Journal* 12: 191–213. https://doi.org/10.1177/1536867X1201200203.

Quintó, L., S. Sanz, E. De Lazzari, and J. J. Aponte. 2012. HTML output in Stata. *Stata Journal* 12: 702–717. https://doi.org/10.1177/1536867X1201200409.

34 DOI: 10.1177/1536867X1701700315

The Stata Journal (2017)
17, Number 3, pp. 774–778

Stata tip 128: Marginal effects in log-transformed models: A trade application

Luca J. Uberti
Department of Politics
University of Otago
Dunedin, New Zealand
luca.jacopo.uberti@gmail.com

Since the introduction of the `margins` command in Stata 11, the empirical literature has increasingly used marginal effects, predictive margins, and adjusted predictions in postestimation analysis. Marginal effects are particularly useful for the interpretation of parameter estimates after `logit`, `probit`, `poisson`, and other nonlinear regression models. If the covariate of interest is in logs, however, obtaining meaningful results from `margins, dydx()` is not straightforward. In this article, I first illustrate these difficulties in the context of estimation with `poisson`. I then suggest that a researcher should always compute the derivative of interest and code it manually with `margins`'s `expression()` option. Lastly, I illustrate these problems using the gravity equation from the trade literature.

I focus on `poisson` for two reasons. First, most examples of `margins` tend to focus on `logit` and `probit`. Second, it is increasingly common for researchers to fit Poisson models with log-transformed covariates. Going beyond traditional applications with count data, some recent literature has shown that `poisson` should be the preferred estimator for constant-elasticity models such as the gravity equation or the Cobb–Douglas production function (Manning and Mullahy 2001; Santos Silva and Tenreyro 2006). In these applications, at least some of the variables on the right-hand side are log transformed.

Let's assume a Poisson model with the following conditional expectation, or "response":

$$E(y_i|x_i) = e^{a+bx_i}$$

y_i and x_i are random variables, b is a vector of estimated parameters, and a is a constant. In Stata language, the "effect" of a covariate x_i is defined as the derivative of the response with respect to x_i:[1]

$$\frac{dE}{dx_i} = be^{a+bx_i}$$

The marginal effect, or average marginal effect (AME), is then the predictive margin (or simply "margin") of this effect. Margins are obtained by averaging or integrating a response over all n observations (or conditions) in the sample (StataCorp 2015, 1404):

$$\text{AME}(x_i) = \frac{1}{n}\sum_{i}^{n} be^{a+bx_i}$$

1. The derivative of the response is itself a response.

In the absence of log transformations, a researcher can obtain this statistic by simply typing `margins, dydx(xi)` after fitting the model (`poisson yi xi`). Things are more complicated, however, when the variable of interest is in logs. If the conditional expectation is specified as $E(y_i|x_i) = e^{a+b\ln x_i}$, the researcher must create a new Stata variable (`gen ln_xi = log(xi)`) before fitting the log-transformed model by typing `poisson yi ln_xi`. The AME then becomes

$$\text{AME}(x_i) = \frac{e^a}{n}\sum_i^n \frac{d}{dx_i}(e^{b\ln x_i}) = \frac{e^a}{n}\sum_i^n bx_i^{b-1} \tag{1}$$

Because `xi` no longer appears in the estimation command, this AME cannot be recovered by typing `margins, dydx(xi)`, which is the case after running the model without log transformations. The command `margins, dydx(ln_xi)`, which Stata does accept, instead computes

$$\text{AME}(\ln x_i) = \frac{e^a}{n}\sum_i^n \frac{d}{d\ln x_i}(e^{b\ln x_i}) = \frac{e^a}{n}\sum_i^n bx_i^b$$

This statistic, however, is not the AME of x_i, nor is it particularly useful for purposes of interpretation.[2]

How can researchers compute the correct AME after fitting models with log transformations? One possible solution is the `margins` command's `expression()` option. In what follows, I illustrate how this option can help researchers interpret the results of constant elasticity models such as the gravity equation. Using trade data from Santos Silva and Tenreyro (2006), I fit a gravity model of trade with form $T_{ij} = gY_i^{\alpha_1}Y_j^{\alpha_2}\tau_{ij}^{\beta}$, where T_{ij} is bilateral trade, Y_i and Y_j are an exporter's and importer's gross domestic product (GDP), respectively, τ_{ij} are bilateral trade costs, and $g = e^G$ is a constant. Trade costs (transportation and tariff costs) are operationalized using geographical distance and an indicator variable that equals one if i and j are members of a free trade agreement. Log-linearizing the gravity equation and exponentiating through, I obtain an expression that may be estimated by `poisson`,

$$T_{ij} = \exp(G + \alpha_1 \texttt{lypex}_i + \alpha_2 \texttt{lypim}_j + \beta_1 \texttt{ldist}_{ij} + \beta_2 \texttt{comfrt}_{ij}) \times \eta_{ij}$$

where `lypex` and `lypim` are, respectively, the log of an exporter's and importer's GDP, `ldist` is the log of geographical distance, `comfrt` is the free trade agreement dummy, and η_{ij} is a multiplicative error term that is uncorrelated with the regressors. The Stata output is displayed below:

2. The same problem arises in an ordinary least-squares context when fitting log-linear or linear-log models.

```
. copy "http://personal.lse.ac.uk/tenreyro/regressors.zip" any_name.zip, public
. unzipfile any_name.zip
    inflating: Log of Gravity.do
    inflating: coding.xls
    inflating: Log of Gravity.dta
    inflating: countrycodes.xls
    inflating: Log of Gravity.xls
successfully unzipped any_name.zip to current directory
. use "Log of Gravity.dta"
(FROM: Santos Silva, J & Tenreyro, S 2006 The log of gravity, RESTAT 88, 641-658)
. poisson trade lypex lypim ldist comfrt, vce(robust) nolog
Poisson regression                              Number of obs     =     18,360
                                                Wald chi2(4)      =    5202.50
                                                Prob > chi2       =     0.0000
Log pseudolikelihood = -1.202e+09               Pseudo R2         =     0.8944

------------------------------------------------------------------------------
             |               Robust
       trade |      Coef.   Std. Err.      z    P>|z|     [95% Conf. Interval]
-------------+----------------------------------------------------------------
       lypex |   .8031383   .0238621    33.66   0.000     .7563694    .8499072
       lypim |   .7982473   .0260628    30.63   0.000     .7471652    .8493294
       ldist |  -.6612203   .0532762   -12.41   0.000    -.7656396   -.5568009
      comfrt |   .3659858   .1738048     2.11   0.035     .0253347    .7066369
       _cons |   -23.2916   1.206315   -19.31   0.000    -25.65594   -20.92727
------------------------------------------------------------------------------
```

The coefficient on `lypex` implies a 0.8% increase in bilateral trade following a 1% increase in country i's GDP. This relation holds at any value of the covariates. Though mathematically simple, this interpretation does not always provide substantively useful insights. The policymaker, for instance, may wish to know exactly by how much we might expect bilateral trade to increase following a spell of rapid economic growth and increased consumer demand in i. To answer this question, we cannot simply type `margins, dydx(lypex)` for the reasons explained above.

The challenge is computing the AME of Y_i:

$$\text{AME}(Y_i) = \frac{1}{n}\sum_{i}^{n}\left(e^{G}\alpha_1 Y_i^{(\alpha_1-1)} Y_j^{\alpha_2}\tau_{ij}^{\beta}\right)$$

Because $\text{AME}(Y_i)$ is a nonlinear combination of both the parameters and the data, the calculation may be accomplished with the `predictnl` command:

```
. predictnl AME = (1/18360)*sum(exp(_b[_cons])*_b[lypex]*(exp(lypex)^(_b[lypex]-1))
> *(exp(lypim)^_b[lypim])*(exp(ldist)^_b[ldist])*exp(_b[comfrt]*comfrt))
. list AME in 18360/18360

        +----------+
        |      AME |
        |----------|
 18360. | 1.56e-06 |
        +----------+
```

This implies that on average, an increase of 1 billion U.S. dollars in i's GDP will lead to an increase in bilateral trade worth 156 million U.S. dollars. We might also wish to know the standard errors of this prediction—and thus whether the trade effects of GDP growth are statistically significant. In principle, the `predictnl` command allows the researcher to specify the `se()` and `p()` options to compute the standard errors and p-values of the prediction. In practice, however, the algorithm of the `predictnl` command may fail to converge. While users may increase the number of iterations with the `iterate()` option, convergence may still not be achieved in large datasets, which is the case in the example above.

A more tractable solution is `margins, expression()`,[3] which allows the explicit specification of a formula for the (derivative of the) response to be margined:

```
. margins, expression(exp(_b[_cons])*_b[lypex]*(exp(lypex)^(_b[lypex]-1)
> *(exp(lypim)^_b[lypim])*(exp(ldist)^_b[ldist])*exp(_b[comfrt]*comfrt))
Warning: expression() does not contain predict() or xb().

Predictive margins                              Number of obs     =     18,360
Model VCE    : Robust

Expression
                 : exp(_b[_cons])*_b[lypex]*(exp(lypex)^(_b[lypex]-1))*
        > (exp(lypim)^_b[lypim])*(exp(ldist)^_b[ldist])*
        > exp(_b[comfrt]*comfrt)
```

	Margin	Delta-method Std. Err.	z	P>|z|	[95% Conf.	Interval]
_cons	1.56e-06	1.24e-07	12.57	0.000	1.32e-06	1.81e-06

In contrast to the syntax for `predictnl`, here we do not need to average the effect (the derivative of the response) by including `(1/18360)*sum()` in the argument of `expression()`. After all, averaging over (some of or all) the covariates is precisely what "taking margins" means.

In sum, the researcher should exercise care when using `margins` to compute the marginal effect of a log-transformed covariate because the `margins` command does not provide an option to exponentiate a logged variable and retrieve the original variable in levels.[4] Thus researchers have no alternative but to calculate the derivative of the response manually and code it into the argument of the `margins` command's `expression()` option. Because manually coding the derivative of the response may quickly become unwieldy in the presence of a large number of variables, the Stata developers should consider updating the `margins` command and including options to exponentiate the variables appearing in the previous regression command. A possible solution might be to allow the argument of `dydx()` to include functions of covariates. For example, it should be possible to compute the AME given by (1) by typing `margins, dydx(exp(ln_xi))`.

3. Another advantage of using `margins, expression()` over `predictnl` is that the former, but not the latter, integrates over unobserved components after models with random effects, for example, `mepoisson` (see [ME] **mepoisson**). This was a new feature in Stata 14.
4. By contrast, the `margins` command does provide users with options to log-transform either the variate (`eydx`) or the covariate of interest (`dyex`), or both (`eyex`).

References

Manning, W. G., and J. Mullahy. 2001. Estimating log models: To transform or not to transform? *Journal of Health Economics* 20: 461–494. https://doi.org/10.1016/S0167-6296(01)00086-8.

Santos Silva, J. M. C., and S. Tenreyro. 2006. The log of gravity. *Review of Economics and Statistics* 88: 641–658. https://doi.org/10.1162/rest.88.4.641.

StataCorp. 2015. *Stata 14 Base Reference Manual*. College Station, TX: Stata Press.

The Stata Journal (2018)
18, Number 1, pp. 287–289 DOI: 10.1177/1536867X1801800117

Stata tip 129: Efficiently processing textual data with Stata's new Unicode features

Alexander Koplenig
Department of Lexical Studies
Institute for the German language (IDS)
Mannheim, Germany
koplenig@ids-mannheim.de

Prior to Stata 13 and especially Stata 14, Stata's abilities to process natural language data were limited because of the string length limit of 244 characters and the lack of Unicode support. To extract basic descriptive information from unformatted text data (for example, word frequency information), one needed to rely on workarounds such as Benoit's (2003) `wordscores` implementation. With Stata 14, this situation has changed. To demonstrate why the new string-processing capabilities of Stata are highly relevant and useful for anyone who deals with natural language data, let us consider, for example, that we want to extract the five most frequent words of the English Universal Declaration of Human Rights. We can do this by first downloading the text file from http://www.unicode.org/udhr/ using the `copy` command. Then, we can use the new string functions `ustrwordcount()` and `ustrword()` to produce language-specific Unicode words that are based on word-boundary rules or dictionaries for languages that do not use spaces between words (for example, for Thai, see below):

```
. copy "http://unicode.org/udhr/d/udhr_eng.txt" udhr.raw, replace
. clear
. local words=ustrwordcount(fileread("udhr.raw"))
. set obs `words'
number of observations (_N) was 0, now 1,963
. generate word=ustrword(fileread("udhr.raw"),_n)
. contract word
. gsort -_freq
. list in 1/5
```

	word	_freq
1.	the	120
2.	and	106
3.	,	95
4.	of	93
5.	to	83

Note that Stata automatically separates punctuation tokens from actual word tokens. In many situations, this is convenient because it makes (effortful) cleaning procedures unnecessary.

In a similar vein, it is easy to extract frequency statistics for n-grams that are sequences of n-consecutive word tokens. Let us say we want to extract the five most frequent word pairs (that is, 2-grams) from the data above. We can do this by generating a word identifier that records the position of each word in the text. The resulting file consisting of all first words of each 2-gram is temporarily stored and then merged with all second words:

```
. generate long wordidentifier=_n
. rename word word1
. tempfile TEMP
. save `TEMP´, replace
(note: file F:\ST_0j000002.tmp not found)
file F:\ST_0j000002.tmp saved
. replace wordidentifier=wordidentifier-1
(1,963 real changes made)
. rename word1 word2
. merge 1:1 wordidentifier using `TEMP´, keep(3)
    Result                           # of obs.
    -----------------------------------------
    not matched                             0
    matched                             1,962  (_merge==3)
    -----------------------------------------

. contract word1 word2
. gsort -_freq
. order word1 word2 _freq
. list in 1/5

     +---------------------------+
     |  word1      word2   _freq |
     |---------------------------|
  1. |      .    Article      30 |
  2. |  right         to      28 |
  3. |    the      right      28 |
  4. |    has        the      25 |
  5. |     of        the      23 |
     +---------------------------+
```

Note that another possibility to extract the most frequent n-grams and corresponding (absolute or relative) frequencies would be to use the `groups` command written by Cox (2017), instead of using the `contract` command before sorting and listing.

As written above, for languages that do not use spaces between words, using the functions `ustrwordcount()` and `ustrword()` has the additional advantage that Stata takes care of the word segmentation. For example, if we want to extract the five most frequent words of the Thai Universal Declaration of Human Rights, we just download the Thai text file. Interestingly, we do not have to change the *loc* argument in the `ustrwordcount()` and `ustrword()` functions. This is because Stata uses the ICU tokenizer, which automatically switches to dictionary-based rules when it identifies particular Unicode script input.

```
. copy "http://unicode.org/udhr/d/udhr_tha.txt" udhr.raw, replace

. clear

. local words=ustrwordcount(fileread("udhr.raw"))

. set obs `words'
number of observations (_N) was 0, now 2,385

. generate word=ustrword(fileread("udhr.raw"),_n)

. contract word

. gsort -_freq

. list in 1/5
```

	word	_freq
1.	และ	107
2.	การ	91
3.	จะ	75
4.	ที่	72
5.	ใน	71

While the ICU documentation (http://userguide.icu-project.org/boundaryanalysis) lists dictionary-based support for Japanese, Khmer, Chinese, and Thai, one can find that other languages, such as Burmese (language-specific ISO 639-3 code: `mya`), Lao (`lao`), or Tibetan (`bod`), are also supported by using the corresponding ISO code in the `copy` command above (`udhr_ISO.txt`).

References

Benoit, K. 2003. Wordscores: Software for coding political texts. http://www.tcd.ie/Political_Science/wordscores/software.html.

Cox, N. J. 2017. Speaking Stata: Tables as lists: The groups command. *Stata Journal* 17: 760–773. https://doi.org/10.1177/1536867X1701700314.

42 DOI: 10.1177/1536867X1801800312

The Stata Journal (2018)
18, Number 3, p. 757

Stata tip 130: 106610 and all that: Date variables that need to be fixed

Nicholas J. Cox
Department of Geography
Durham University
Durham, UK
n.j.cox@durham.ac.uk

On Statalist and elsewhere, people sometimes try to work with monthly date variables with values like 195201 or 201805. You get the idea: 195201 is January 1952 and 201805 is May 2018. The advantages of such a representation are twofold. People can quickly grasp the convention. Such dates sort correctly into chronological order. (For that to work, 01 to 09, rather than 1 to 9, are essential codings for January to September.) Despite these advantages, such dates are useless for any other serious statistical or Stata purpose. The point of this tip is to explain precisely why that is so and then what else you should do.

There is no homunculus or other intelligence inside Stata that sees such a variable and thinks "Oh! A monthly date". Nor will trying to apply a monthly date format help. For more on why changing the display format does not work here, see Cox (2012). People trying that typically realize that they need something else. People not trying that often get stuck.

To see the main problem, focus on what happens at the turn of each calendar year, say, as 201712 (December 2017) turns into 201801 (January 2018). You understood in reading that example that 201801 follows 201712 immediately and that there were no months that would be 201713 to 201799. On that point, and many others, you are better informed and smarter than Stata. Stata can see only a numeric gap of 89, because $201801 - 201712 = 89$. Hence, such dates show, through any 12-month period, 11 steps or gaps of 1 as you proceed from January to December and 1 big step or gap of 89 as you proceed from December to January. That big gap will mess up almost anything graphical or statistical using the date variable. In particular, graphs will look crazy and `tsset` or `xtset` in terms of the time variable will also flag (but not correct for) uneven spacing of your data, which will mess up any modeling or calculation using lags or leads that depend on either setting.

There are various ways to map such variables to Stata monthly dates as usually understood. Such dates have origin (that is, are 0) at January 1960. Typically, a monthly date format is applied so that people see monthly dates they can understand. Let's imagine a toy dataset with those two run-together monthly dates as values.

```
. clear
. input bad_month_date
        bad_mon~e
  1. 195201
  2. 201805
  3. end
```

The best way to process such a numeric variable is to split values into year and month components and then feed those as arguments to `ym()`, a function that expects numeric year and month values. The year is the value divided by 100 and rounded down, and the month is the remainder after dividing by 100:

```
. generate good_month_date =
> ym(floor(bad_month_date/100), mod(bad_month_date, 100))
. format good_month_date %tm
. list

     +-----------------------+
     | bad_mo~e   good_m~e   |
     |-----------------------|
  1. |   195201     1952m1   |
  2. |   201805     2018m5   |
     +-----------------------+
```

Here we just applied the default monthly date format. You can apply many other formats, but see the help for `datetime display formats` for the details. For more on `floor()` or `mod()` if you want it, see their respective help files or Cox (2003, 2007).

There are other ways to do that conversion. You might have thought of converting the date to string, extracting year and month components using `substr()`, and then passing the results back through `real()` and then `ym()`. Those types of conversions back and forth can be avoided, as just explained, by keeping the conversion as an entirely numeric operation. Note that at the time of writing (Stata 15.1), the nested function call `monthly(string(), "YM")` is not yet smart enough to parse run-together monthly dates.

Let's complete the circle and imagine that for some purpose, bizarre or other, you need to export monthly dates in the form 195201 or 201805. A good reason for that would be if some other software understands (or even requires) such a format. If you have a variable like `bad_month_date`, you already have what you need. If you have a variable like `good_month_date`, then push it through `string(good_month_date, "%tmCYN")`.

It follows from this that daily dates that have the form 20180531 have the same advantages and disadvantages. In practice, people seem to realize more quickly that they need to convert such dates to daily dates as understood by Stata. The example

```
. display %td daily("20180531", "YMD")
31may2018
```

shows that first converting the date to string and then feeding the result to `daily()` will work fine. You need to follow with assignment of a daily date format.

Here is an explanation for anyone who wants it of the title of this tip. The Battle of Hastings—a key event in British (even more parochially, English) history—was in October 1066. That year is perhaps now remembered more for an irreverent send-up of that history, or rather how it was once taught (Sellar and Yeatman 1930). See also, or not as the mood takes you, Cox (2011).

References

Cox, N. J. 2003. Stata tip 2: Building with floors and ceilings. *Stata Journal* 3: 446–447. https://doi.org/10.1177/1536867X0400300413.

———. 2007. Stata tip 43: Remainders, selections, sequences, extractions: Uses of the modulus. *Stata Journal* 7: 143–145. https://doi.org/10.1177/1536867X0700700113.

———. 2011. Speaking Stata: MMXI and all that: Handling Roman numerals within Stata. *Stata Journal* 11: 126–142. https://doi.org/10.1177/1536867X1101100109.

———. 2012. Stata tip 113: Changing a variable's format: What it does and does not mean. *Stata Journal* 12: 761–764. https://doi.org/10.1177/1536867X1201200415.

Sellar, W. C., and R. J. Yeatman. 1930. *1066 and All That: A Memorable History of England*. London: Methuen.

The Stata Journal (2019)
19, Number 3, pp. 738–740 DOI: 10.1177/1536867X19874263

Stata tip 131: Custom legends for graphs that use translucency

Tim P. Morris
MRC Clinical Trials Unit
University College London
London, UK
tim.morris@ucl.ac.uk

As of version 15, Stata graphs permit translucent elements, which are invoked by "%#" following a *colorstyle* (where # is a percentage of opacity). Many `twoway` commands will include a legend. This gives control over the labels but not over the symbols.

I will now demonstrate a trick to produce legend symbols "by hand" for plot types `scatter` and `line`, and I will demonstrate why the arrival of translucency has prompted the need for this. This tip is not well showcased by the `sj` graph scheme; hence, I advise readers to try running the examples using a scheme including colors.

First, we will simulate a dataset with two groups, each containing 1,000 observations per group on `y` and `x` (`drawnorm` sets the dataset size from empty).

```
. clear
. matrix means = (4,1)
. matrix corr = (1, .5  .5, 1)
. drawnorm y1 x1, n(1000) seed(1)
. drawnorm y2 x2, corr(corr) means(means)
```

We begin with a scatterplot:

```
. twoway
>  (scatter y1 x1, mcolor(%30) mlcolor(white%1))
>  (scatter y2 x2, mcolor(%30) mlcolor(white%1))
>  ,
>  legend(order(1 "Group 1 (default symbol)" 2 "Group 2 (default symbol)"))
```

The scatterplot above includes translucent symbols with 30% opacity. The translucency inherited by the symbols in the legend is not necessarily desirable, because it matches the palest possible shade seen within the plot, which may make it harder to match the legend to the plotted data. To produce a legend symbol of our own choosing, we can add what I term a "ghost" plot for each symbol we wish to control. This is achieved as follows:

```
. twoway (scatteri . . , msize(medlarge))
>  (scatteri . . , msize(medlarge))
>  (scatter y1 x1, pstyle(p1) mcolor(%30) mlcolor(white%1))
>  (scatter y2 x2, pstyle(p2) mcolor(%30) mlcolor(white%1))
>  , legend(order(1 " Group 1 (custom symbol)" 2 " Group 2 (custom symbol)"))
```

Figure 1 shows the plot with the first and second versions of the legend, placed in one graph here to emphasize the difference: the top row uses the default translucent symbols, and the bottom row uses the opaque custom symbols produced by the ghost plots.

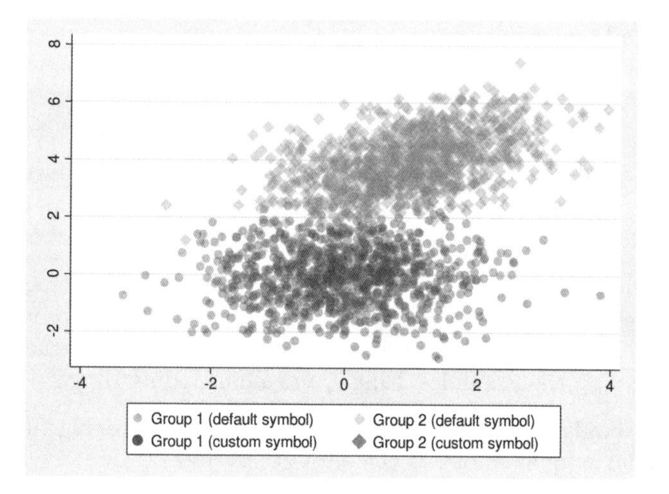

Figure 1. Scatterplot of simulated data with default (upper row) and custom (lower row) legend symbols

The ghost plots here were invoked using `scatteri`, and the points `. .` ensured nothing was plotted—but the symbol attributes nonetheless appeared in the legend. The option `legend(order(1 "..." 2 "..."))` now refers to our ghost plots (because they appear first and second in the `twoway` command). Our original plots are now in positions 3 and 4, meaning that the symbols they use would have changed. This is remedied by the option `pstyle(p1)` for plot 3 and `pstyle(p2)` for plot 4.

To demonstrate this method using a `line` plot type, we first (twice) `reshape` the simulated data:

```
. generate int id = _n
. reshape long y x, i(id) j(group)
. rename y y1
. rename x y0
. reshape long y, i(id group) j(time)
```

The following `twoway` command then produces a legend with user-manipulated symbols:

```
. twoway
>  (function . , lcolor(gs10) lwidth(medthick))
>  (function . , lcolor(gs6)  lwidth(medthick))
>  (line y time if group==1, pstyle(p1line) lcolor(gs10%5))
>  (line y time if group==2, pstyle(p2line) lcolor(gs6%5))
>  , legend(order(1 "Group 1 (custom symbol)" 2 "Group 2 (custom symbol)") cols(1))
```

Most of this works in the same way as with `scatter`. The ghost plot type is `function`, and again we use it to plot nothing (`.` denotes missing). The *pstyles* used by plots 3 and 4 are now `p1line` and `p2line`, respectively.

Using this approach, any options required for the legend can be achieved by manipulating options within the ghost plots, and this manipulation does not have to affect the "living" plots.

The trick outlined in this tip has uses beyond translucency. In presentations, it is sometimes desirable to add elements gradually to build up a graph. For example, the legend that comes with `msymbol(p)` is almost never readable, even when the color of a swarm of points is. Another use is in presentations. One may begin with the graph region with no data but a legend, then introduce the data for group 1 and subsequently group 2, etc.; using this trick, the legend could be present from the beginning.

DOI: 10.1177/1536867X19874264

The Stata Journal (2019)
19, Number 3, pp. 741–747

Stata tip 132: Tiny tricks and tips on ticks

Nicholas J. Cox
Department of Geography
Durham University
Durham, UK
n.j.cox@durham.ac.uk

Vince Wiggins
StataCorp
College Station, TX
vwiggins@stata.com

1 Ticks are helpful, but how can we tune them?

Ticks on graph axes are small but ideally useful details that you may often want to tune. This column is a collection of small points on those small ticks, intended to explain what is not quite obvious—or not quite obviously useful—on reading the documentation. Naturally and necessarily, the topic has implications for axis labels too.

2 Removing ticks, or the charms of invisibility and nonexistence

Ticks classically mark steps on a continuous numerical scale, just like marks on a ruler showing lengths in centimeters and millimeters or inches and fractions of an inch. If you are as old as or even older than the authors, you drew your first graphs in school on squared paper, hoping not to make too many silly errors or smudges or splodges. You drew axes and then annotated them with ticks and labeled some if not all of the ticks. Now we have commands like `graph` to do it, but their defaults may not give you exactly what you want.

The first tip is that sometimes the ticks can be removed without pain, especially if they do not help or are misleading because the items on an axis are distinct categories, not values on a numerical scale. Thus, in comparing mileage for foreign and domestic cars using a `dotplot`, say,

```
. sysuse auto
. dotplot mpg, over(foreign) ylabel(, angle(horizontal))
```

ticks appear against "Domestic" and "Foreign" (figure 1). Tufte's term "chartjunk" (Tufte 2001) is a little strong for this case, but the ticks convey or imply nothing we do not know otherwise and so could be deleted.

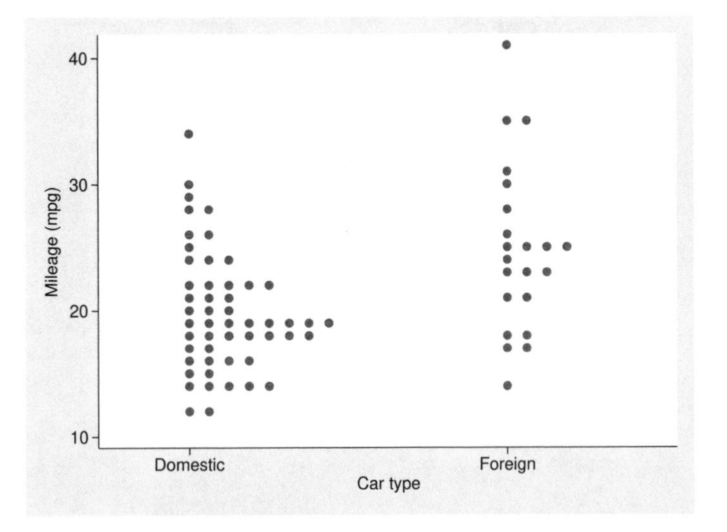

Figure 1. Dotplot of mileage by domestic and foreign cars. Do we need the ticks on the x axis?

What is going on? `dotplot` is just a wrapper for `scatter`, but `scatter` neither knows nor cares exactly what kind of numeric variable is being fed to it. Further, `dotplot` is broad-minded enough to accept numeric and string variables as arguments to `over()` without wanting to decide for you what kind of variable you have and in particular whether ticks make sense. Implicit there is the fact that `dotplot` maps your input to always produce a numeric x-axis variable inside the command.

To get rid of the ticks, you can just add the option

```
xlabel(, noticks)
```

and they disappear. But now you may complain that the labels are too close to the axis. If Stata could talk back, which the company should be working on for a future version, it would be a little sharp and say that it gave you precisely what you asked for. But there is another way. You can make the tick invisible with

```
xlabel(, tlcolor(bg))
```

because changing the color to `bg` (background) suffices to make the tick invisible. The results of `xlabel(, noticks)` and `xlabel(, tlcolor(bg))` are shown side by side in figure 2.

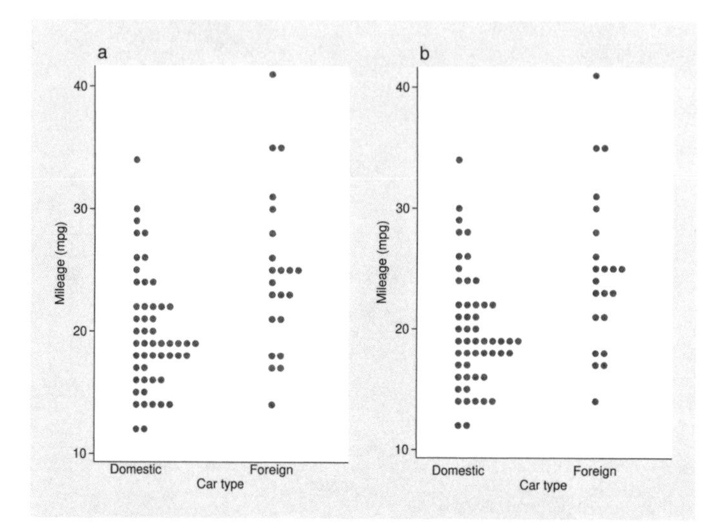

Figure 2. Two different ways of removing the ticks in figure 1. In a, the ticks really are removed. In b, the tick line color is set to the same as the background color, which leaves the label text exactly where it was.

Let us underline a monumentally noteworthy principle: if you cannot see it, that is psychologically equivalent to it not being there. The cosmological, philosophical, and theological versions of this principle are best reserved for other accounts.

But still you might complain. We have removed the ticks and kept the text labels where they were, but now without their ticks, the text labels appear displaced to the left. The next trick here is to put the text labels on a very small slant. `graph` will then flip the text. First, the code respects the sign (not sine!) of the angle—whether zero, negative, positive—in aligning text so that it is centered, or flush-left, or flush-right, given the tick position. The code also respects the angle (or, if you like, the sine) in deciding on the slant.

```
. dotplot mpg, over(foreign) ylabel(, angle(horizontal))
> xlabel(, angle(-0.001) tlcolor(bg))
```

Using −0.001 is a fudge, but if it looks as you wish, then that's what counts; you will be happy, and your readers should be too. Please look ahead to figure 3a for the result.

Obvious, but worth a little emphasis: it is best to have text in graphs aligned horizontally for ease of reading (Tufte [2001] again, if you want a reference). `graph` lets you, for example, put axis labels at, say, 45 degrees, which Stata people often exploit, but that is still a device of despair. The pejorative "giraffe graphics" has been applied to graphics that require willingness to hold the reader's head at arbitrary angles (Cox 2004).

Now again the text is where you want, but possibly again too close to the axis. We have yet further dodges. We can increase the length of our still invisible tick.

```
. dotplot mpg, over(foreign) ylabel(, angle(horizontal))
> xlabel(, angle(-0.001) tlength(*2) tlcolor(bg))
```

Figure 3 shows the two steps side by side.

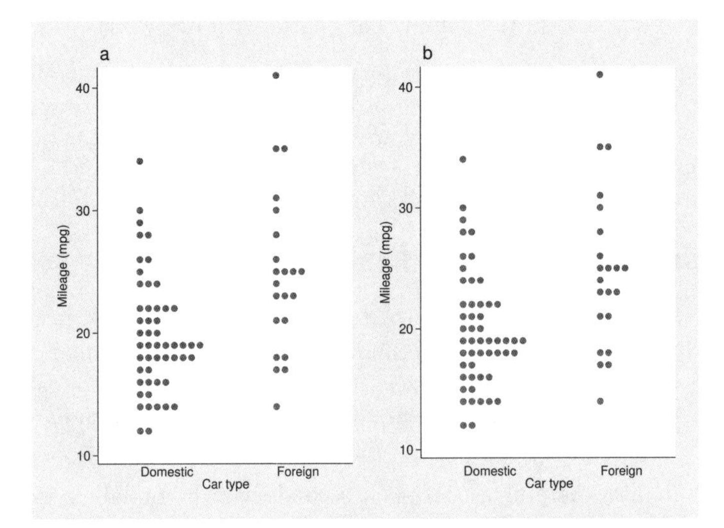

Figure 3. Two steps in aligning axis label text. In a, cunning choice of angle puts text underneath each dot pattern. In b, the invisible tick has been lengthened to push the text away from the axis.

There is an easy corollary: as tick length can be tuned, so also `tlength(0)` is a further way to remove ticks. So we now know three ways to remove ticks: the suboptions `noticks`, `tlcolor(bg)`, and `tlength(0)`. That is not too many ways to do it. Sometimes, we want to combine those suboptions, as we have already seen. Note for your future graphing: if one way does not work because of a program quirk or even a bug, try using another way. (We have not yet touched on `tlstyle()` or `tlwidth()`, with which keen readers may wish to experiment.)

3 Adding ticks where none are shown

Quite the opposite issue can arise with `graph bar` or `graph hbar`—and occasionally also with `graph box`, `graph hbox`, or `graph dot`. Underlying the design and implementation of these commands is the idea that one axis is to be thought of as categorical, so ticks on that axis are not the default and not easily added either.

A common example: people want a bar chart of something that varies with time, say, over a period of years, and seek more control over the time axis than is given or allowed by `graph bar`. Those wanting this rarely use `graph hbar`, given strong conventions in many fields that time belongs on the horizontal axis, but it would pose the same problems.

Perhaps your years are irregularly spaced, but `graph bar` ignores the gaps. It just puts the categories it sees (years to you) in order. Or you have so many years that the labels run into each other, and you would be happy with labels every other year or every fifth year. Or you wish to combine bar charts with line charts.

Instead of giving detailed tricks for this situation, we confine ourselves to a broad strategic hint. The best way to advance in this circumstance is to retreat. You would be better off with `twoway bar`. Then your time axis (x axis) is numeric, and the rest of `twoway` is there for the asking.

4 Two kinds of ticks, and labels too

A feature often overlooked is that you can have different kinds of ticks (and labels) on each axis that allows them. The secondary labels are called minor, although that describes only their default smaller sizes. You can make the minor labels and ticks bigger, longer, or in other ways different from the default by reaching in and tuning suboptions.

A simple but often helpful use of such secondary ticks or labels is to emphasize individual crucial values: thresholds for action, major events or stages, or whatever.

We illustrate with data on successive records for the number of decimal places of the constant π known correctly at different times. The dataset used is downloadable from the resources for this issue at the *Stata Journal* website. The data were read from https://en.wikipedia.org/wiki/Chronology_of_computation_of_pi on April 18, 2019. The `notes` for the dataset carry some qualifications. The graph to come was inspired by one given by Trefethen (2011, 277).

Because this dataset will become rapidly outdated, note that at the time of writing the record number of decimal places known correctly was as reported on March 14, 2019, by Emma Haruka Iwao: 31,415,926,535,897 digits, although as yet there are no media reports of anyone memorizing them all.

A plot of these data requires (nay, demands) a logarithmic scale for number of decimal places. Whether time should be plotted differently we leave as an open question (Cox 2012). The plot here (figure 4) focuses only on records since 1800. As is evident, a crucial shift in behavior was the first use of digital computers for this purpose in 1949. The code shown for the graph should be almost self-explanatory. Given the very large numbers to be shown, the axis labels are just powers of 10. After reading the dataset, we use `clonevar` to ensure copying of the variable label but promptly `replace` with the logarithm to base 10.

```
. use pi_decplaces, clear
. clonevar log_places = places
. replace log_places = log10(places)
. twoway connected log_places year if year > 1800,
> ylabel(0(1)13, angle(horizontal))
> ms(o) xlabel(1844 1879 1914 1984 2019)
> xmlabel(1949, labsize(*2) tlength(medlarge))
> xtitle("") subtitle(powers of 10, place(w))
```

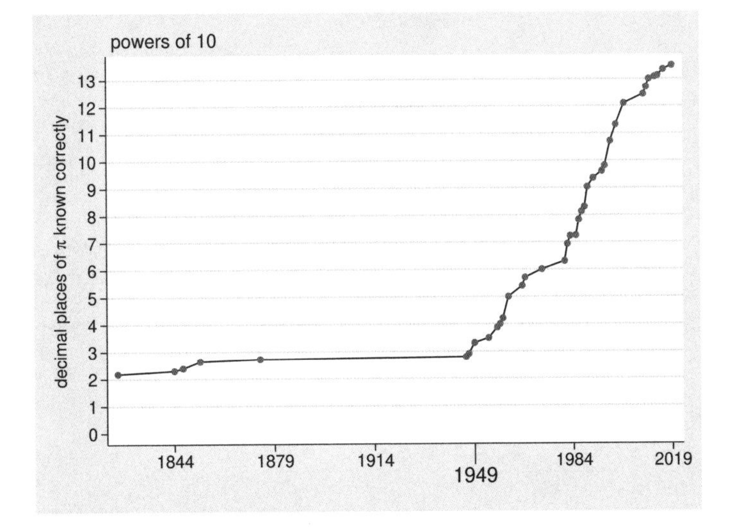

Figure 4. Although this plot is intriguing and inspiring in its own right, here it is a vehicle to show the use of secondary labels to flag a key date, 1949, as the first use of digital computers to calculate the decimal digits of π.

5 Columns and tips with further tips

We here cross-reference other discussions in previous issues of the *Stata Journal*.

Suppose your data span a relatively small number of years (or other time intervals). Hence, each interval can be labeled, and each year accounts for an appreciable fraction of the time axis. The ticks would be better placed at the end of each interval, which is the beginning of the next interval. Text labels with no ticks then belong in the middle of each interval. See Cox (2007) for the details.

Suppose you have chosen a nonlinear scale (logarithmic or other, say, square root or logistic) and wish to see "nice" numbers on your scale, which thus will not be equally spaced on a graph axis. See Cox (2018) for logarithmic scales and Cox (2008, 2012) more generally.

References

Cox, N. J. 2004. Speaking Stata: Graphing categorical and compositional data. *Stata Journal* 4: 190–215. https://doi.org/10.1177/1536867X0400400209.

———. 2007. Stata tip 55: Better axis labeling for time points and time intervals. *Stata Journal* 7: 590–592. https://doi.org/10.1177/1536867X0800700410.

———. 2008. Stata tip 59: Plotting on any transformed scale. *Stata Journal* 8: 142–145. https://doi.org/10.1177/1536867X0800800113.

———. 2012. Speaking Stata: Transforming the time axis. *Stata Journal* 12: 332–341. https://doi.org/10.1177/1536867X1201200210.

———. 2018. Speaking Stata: Logarithmic binning and labeling. *Stata Journal* 18: 262–286. https://doi.org/10.1177/1536867X1801800116.

Trefethen, L. N. 2011. *Trefethen's Index Cards: Forty Years of Notes about People, Words, and Mathematics.* Singapore: World Scientific. https://doi.org/10.1142/8207.

Tufte, E. R. 2001. *The Visual Display of Quantitative Information.* 2nd ed. Cheshire, CT: Graphics Press.

The Stata Journal (2019)
19, Number 4, pp. 1009–1014 DOI: 10.1177/1536867X19893643

Stata tip 133: Box plots that show median and quartiles only

Nicholas J. Cox
Department of Geography
Durham University
Durham, UK
n.j.cox@durham.ac.uk

1 Introduction

Box plots (for example, Tukey [1977]) are well known as summary plots for univariate distributions. In the most common design, as supported by `graph box` and `graph hbox`, a central box for each variable or group shown on a graph depicts median and lower and upper quartiles. The length of the box is thus the difference or distance

$$\text{upper quartile} - \text{lower quartile}$$

known as the interquartile range. What is shown beyond the box may include individual data points if any lie more than 1.5 times the interquartile range from the nearer quartile. Otherwise, capped lines known as whiskers are shown; these extend to the outermost data points not shown individually.

A detail of importance to what follows is that no whiskers are shown if no value is less than the lower quartile or more than the upper quartile. That can happen with data, especially with very small samples or variables showing many ties, such as grades (ordered responses coded 1 to 5, or whatever), counts, or integer scores.

Requests are often made for Stata code to produce truncated box plots that show only median and quartiles. This tip shows how to get such plots without elaborate programming.

2 Median and quartiles only

Wanting only median and quartiles echoes a very common twofold practice:

- Show a summary of level (say, a mean) of some variable or group by a marker, in this context most often a filled circle.
- Indicate variability by capped or uncapped spikes extending above and below the marker for level by some multiple of the standard deviation of the data or the standard error of the mean.

Such plots thus include, but are not restricted to, displays of confidence intervals. Wanting plots that show median and quartiles is natural whenever there is interest or use in summaries that are more robust or resistant than the mean and the standard deviation, or summaries dependent on those statistics. For example, the mean, and even more the standard deviation, can be highly sensitive to outliers, especially in small samples. Wanting to use box plot conventions for showing median and quartiles also reduces the scope for misreading such displays as based on means and standard deviations or standard errors.

Firing up a typical box plot, say, with

```
. sysuse auto
. graph box mpg, over(foreign) ylabel(, angle(horizontal))
```

may suggest that a way to proceed is just to make the whiskers and individual points invisible. We could add further options to do that:

```
. graph box mpg, over(foreign) ylabel(, angle(horizontal)) cwhiskers
> lines(lcolor(none)) marker(1, ms(none))
```

This trick is always worth knowing, but it is not a perfect solution. If there are two or more kinds of markers, all must be removed. Further, and more serious in practice, such a technique makes elements invisible without removing the space needed to show them. At worst, the result is that the boxes occupy just a small fraction of the plot region, which is not usually what is wanted. The same comments apply to removing or hiding elements in the Graph Editor.

Elsewhere (Cox 2009, 2013) there is detailed discussion of how you can make your own box plots, or variations on them, by first calculating the summary statistics you want to show and then calling up `graph twoway`. This approach is necessary if you want to show something quite far removed from what is offered by `graph box` or `graph hbox`. Examples might be showing means as well as medians and quartiles or using customized whiskers that extend to particular percentiles (for example, 5% and 95%) or the sample extremes.

3 Collapsing to samples of three

The main focus of this tip is another method that is flexible, easy to understand, and easy to implement. Suppose we reduce each group or variable to three values that are the median and quartiles. Then if such values are now the data, with samples of size 3, Stata's rules imply that the smallest and largest values are taken as the quartiles and the middlemost (a splendid word for the one in the middle) is taken as the median. Thus, a box plot will then show the median and quartiles of the original data, and only those, with no whiskers or individual data points. Skip or skim ahead to (*) if you are unsurprised by that or happy to believe that there is a detailed explanation.

It is standard that the median of any odd number of values, including three, is the middlemost. What may be a little more surprising, or at least unexpected, is the rule for quartiles. Stata's recipe is documented at [R] **summarize**.

For sample size n, we consider values x ordered such that $x_{(1)} \leq \cdots \leq x_{(n)}$. For the pth percentile, we apply weights $w_{(i)}$ scaled to sum to n and cumulate $P = \sum_{j=1}^{i} w_i$. `summarize` reports as percentile

$$(x_{(i-1)} + x_{(i)})/2 \quad \text{if} \quad P = np/100$$

and $x_{(i)}$ otherwise.

For $n = 3$, $p = 25$, and $np/100 = 75/100$. With equal weights $w_i = 1$, P runs 1, 2, and 3 for three ordered values, so we use $x_{(1)}$, the lowest value, as the lower quartile. Similarly, for $n = 3$, $p = 75$, $np/100 = 225/100$, so we use $x_{(3)}$, the largest value, as the upper quartile.

The case of unequal weights is not relevant to this point. We need to know only that if presented with three values, Stata's box plot commands will echo precisely those values as median and quartiles. It is not contradictory that those three values could have been produced by a calculation using weights of any kind.

These rules naturally do not exclude the possibility that two or even three of those summaries coincide numerically, finally producing a degenerate box in a box plot.

(∗) Reduction to sets or subsets of three values can be achieved by an easy `collapse` and `reshape` of the data. The dataset in memory will thus be destroyed. Hence, it should be `save`d first or beforehand. Or you should use `preserve` before all that follows, and `restore` afterward, to get the original dataset back in memory when graphing is finished. See the help files for these commands if the idea is new to you.

Before we do this, there is one small warning. On a `collapse`, variable labels you want to be used as axis titles in the graph may disappear. If they are long and complicated, you may find it irritating to have to type them again. One way to store them safely is to put them in local macros. Continuing our example with the auto data, let's type

```
. local ytitle: var label mpg
. local xtitle: var label foreign
```

Note that we do not need to type the variable label itself. Stata knows what it is, and the so-called extended macro function `: var label` is the hook to retrieve it. This is particularly useful if the variable label is long or complicated, say, with any or all of superscripts, subscripts, Greek letters, Stata Markup and Control Language, or other unusual characters. `help extended macro functions` gives you access to documentation on this and several other useful hooks.

Now to the nub of the matter. The call to `collapse` will be a single line, such as

```
. collapse (p25) mpg25=mpg (p50) mpg50=mpg (p75) mpg75=mpg, by(foreign)
```

For one variable, we need to ask for median and quartiles of that variable, here as percentiles for 25%, 50%, and 75%. We often want to do that separately by distinct combinations of one or more variables. As far as `collapse` is concerned, the new variables containing median and quartiles could be called anything legal, such as `frog`, `toad`, and `newt`. Using the same prefix (here `mpg`) and different numeric suffixes looks ahead to what will be easiest when we `reshape`. If we overlook that, we will likely just have to `rename` some variables.

Let's see what that produces and what a `reshape` produces in turn.

```
. list

     +---------------------------------------+
     |  foreign   mpg25    mpg50    mpg75    |
     |---------------------------------------|
  1. | Domestic    16.5       19       22    |
  2. |  Foreign      21     24.5       28    |
     +---------------------------------------+

. reshape long mpg, i(foreign) j(pctile)
(note: j = 25 50 75)
Data                                   wide   ->   long
-----------------------------------------------------------------------------
Number of obs.                            2   ->      6
Number of variables                       4   ->      3
j variable (3 values)                         ->   pctile
xij variables:
                          mpg25 mpg50 mpg75   ->   mpg
-----------------------------------------------------------------------------

. list, sepby(foreign)

     +----------------------------+
     |  foreign   pctile     mpg  |
     |----------------------------|
  1. | Domestic       25    16.5  |
  2. | Domestic       50      19  |
  3. | Domestic       75      22  |
     |----------------------------|
  4. |  Foreign       25      21  |
  5. |  Foreign       50    24.5  |
  6. |  Foreign       75      28  |
     +----------------------------+
```

Now we can draw our graph (figure 1):

```
. graph box mpg, over(foreign) ylabel(, angle(horizontal)) ytitle("`ytitle'")
> b1title("`xtitle'")
```

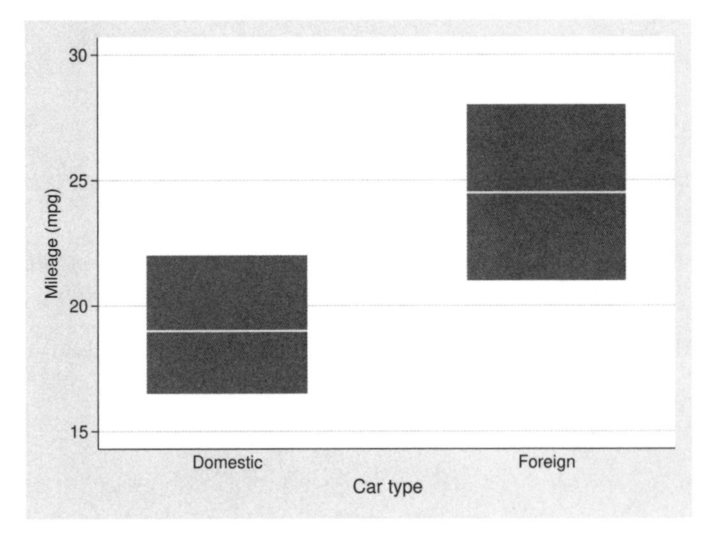

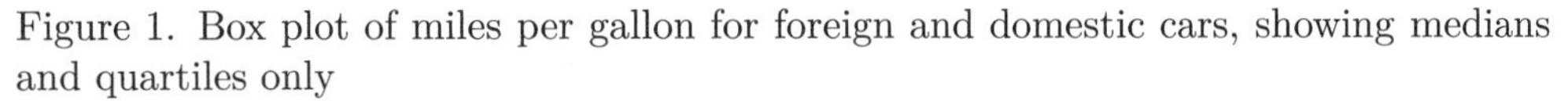

Figure 1. Box plot of miles per gallon for foreign and domestic cars, showing medians and quartiles only

4 Complications: More variables, more categories, weights, ...

For more variables, we will need to work a bit more in specifying what is wanted to `collapse`. This example with temperatures for cities in the United States is reproducible, but the graph is not shown for reasons of space. The vertical axis labels are a nod to readers in most countries of the world more accustomed to Celsius temperatures: $32°\text{F} = 0°\text{C}$, $50°\text{F} = 10°\text{C}$, and so forth.

```
. sysuse citytemp, clear
. collapse (p25) Jan25=tempjan Jul25=tempjuly (p50) Jan50=tempjan
> Jul50=tempjuly (p75) Jan75=tempjan  Jul75=tempjul, by(region)
. reshape long Jan Jul, i(region) j(pctile)
. label var Jan "January"
. label var Jul "July"
. graph box Jan Jul, over(region) ytitle(Mean temperature (&degreeF))
> ylabel(, angle(horizontal)) yline(32) ylabel(14(18)86)
```

More variables as categorical classifiers? Just feed them all to the `by()` option of `collapse` and the `i()` option of `reshape`.

Weights? Spell them out to `collapse`. We do not need to worry about them beyond that point because the graphical problem is to deal with our already weighted medians and quartiles.

Anyone doing this often would be well advised to write a do-file or command that automates the process, but that goes further than the aim of this tip.

References

Cox, N. J. 2009. Speaking Stata: Creating and varying box plots. *Stata Journal* 9: 478–496. https://doi.org/10.1177/1536867X0900900309.

———. 2013. Speaking Stata: Creating and varying box plots: Correction. *Stata Journal* 13: 398–400. https://doi.org/10.1177/1536867X1301300214.

Tukey, J. W. 1977. *Exploratory Data Analysis.* Reading, MA: Addison–Wesley.

The Stata Journal (2019)
19, Number 4, pp. 1015–1020 DOI: 10.1177/1536867X19893644

Stata tip 134: Multiplicative and marginal interaction effects in nonlinear models

William H. Dow
University of California, Berkeley
Berkeley, CA
and National Bureau of Economic Research
Cambridge, MA
wdow@berkeley.edu

Edward C. Norton
University of Michigan
Ann Arbor, MI
and National Bureau of Economic Research
Cambridge, MA
ecnorton@umich.edu

J. Travis Donahoe
Harvard University
Cambridge, MA
jtdonahoe@g.harvard.edu

1 Introduction

In Stata tip 87, Buis (2010) demonstrated how to use Stata to calculate multiplicative interaction effects in nonlinear models. As Buis notes, multiplicative interaction effects, such as odds ratios from a `logit` model (see [R] **logit**), are often easily obtained from standard Stata output without additional programming. Buis contrasts this with marginal interaction effects, which require additional postregression programming for correct computation (Ai and Norton 2003)—although the `margins` commands in Stata have greatly simplified this computation and can produce correct marginal effect calculations with only a few lines of additional postregression commands (Karaca-Mandic, Norton, and Dowd 2012). Buis also explains that multiplicative and marginal interaction effects each answer different questions; thus, it is important for analysts to have both in their toolkit.

Nevertheless, we have observed numerous authors misinterpret Buis (2010), citing the article as justification for presenting only multiplicative interaction effects, claiming they are easier to calculate or interpret. For example, Doidge, Karolyi, and Stulz (2013) state, "We, therefore, report the regression coefficients, but interpret them in terms of odds ratios which are simpler to interpret when there are interaction terms in the model (see, for example, Buis, 2010; Kolasinski and Siegel, 2010)." Similarly, Vaidyanathan (2011) states, "While scholars continue to debate how to interpret interaction effects in nonlinear models, Buis 2010 argues that using multiplicative effects, such as odds ratios, overcomes most difficulties, and I follow him in this regard."

In this Stata tip, we present a simple stylized example illustrating the starkly different conclusions that marginal and multiplicative interaction effects can imply. We argue that unless analysts have a strong theoretical preference, they should routinely calculate and present both marginal and multiplicative interaction effects after fitting nonlinear models.

2 Multiplicative and marginal interaction effects on probability and odds scales

For illustration, we focus on the `logit` estimator with a binary outcome variable y, which is modeled as a function of two interacted binary explanatory variables, x and z. In other words, $y = f(\beta_1 x + \beta_2 z + \beta_{12} x \times z)$. However, the implications of this article directly generalize to all nonlinear models, including count data models such as Poisson (see [R] **poisson**) or negative binomial (see [R] **nbreg**), which are often interpreted multiplicatively.

We define and compare four ways of representing an interaction effect after fitting a `logit` model. These include multiplicative and marginal interaction effects on both the probability scale and the odds scale. Define p as the predicted probability for a binary dependent variable y, conditional on values of binary x and z and their interaction. The multiplicative interaction effect of changes in both x and z on p can be represented on the probability scale as

$$\frac{p_{x=1,z=1}/p_{x=1,z=0}}{p_{x=0,z=1}/p_{x=0,z=0}}$$

One can easily calculate this in Stata using the `nlcom` command after `margins` following the logistic regression:

```
. logit y i.x##i.z
. margins x#z, post
. nlcom (_b[1.x#1.z] / _b[1.x#0.z]) / (_b[0.x#1.z] / _b[0.x#0.z])
```

In contrast with the above multiplicative interaction effects, marginal interaction effects are represented as a difference in differences on the probability scale:

$$(p_{x=1,z=1} - p_{x=1,z=0}) - (p_{x=0,z=1} - p_{x=0,z=0})$$

One can calculate this marginal interaction effect on the probability scale using a one-line postestimation command:

```
. margins, dydx(z) at(x=(0 1)) contrast(atcontrast(r._at)) post
```

It is also common in some fields to present effects on the odds $p/(1-p)$ scale rather than probability scale p. If p is close to zero, then the results on both scales are usually similar, but more generally they may differ. The multiplicative interaction effect on the odds scale is

$$\frac{\frac{p_{x=1,z=1}}{1-p_{x=1,z=1}} \Big/ \frac{p_{x=1,z=0}}{1-p_{x=1,z=0}}}{\frac{p_{x=0,z=1}}{1-p_{x=0,z=1}} \Big/ \frac{p_{x=0,z=0}}{1-p_{x=0,z=0}}}$$

This is even simpler to calculate in Stata using the `or` option in the `logit` command, which is one reason why it is among the most commonly reported variants:

```
. logit y i.x##i.z, or
```

Finally, although rarely used, it is also possible to calculate marginal interaction effects on an odds scale (this is the marginal effect variant presented in Buis [2010]):

$$\left(\frac{p_{x=1,z=1}}{1-p_{x=1,z=1}} - \frac{p_{x=1,z=0}}{1-p_{x=1,z=0}}\right) - \left(\frac{p_{x=0,z=1}}{1-p_{x=0,z=1}} - \frac{p_{x=0,z=0}}{1-p_{x=0,z=0}}\right)$$

```
. margins x#z, expression(exp(xb())) post
. lincom (_b[1.x#1.z] - _b[1.x#0.z]) - (_b[0.x#1.z] - _b[0.x#0.z])
```

Similar Stata code can be written to calculate interaction effects after probit, Poisson, and negative binomial models.

3 A simple cautionary example

To illustrate the potential danger of reporting only one variant of the above interaction effect calculations, we present an example in which each of these variants implies different conclusions. Our simple example has just four data points (see panel A of figure 1). When $x = 0$, as z increases from 0 to 1, the probability p of a positive outcome rises from 0.05 to 0.10. When $x = 1$, as z increases from 0 to 1, the probability of a positive outcome rises from 0.10 to 0.19. Panel B of figure 1 shows the same four data points transformed to the odds scale.

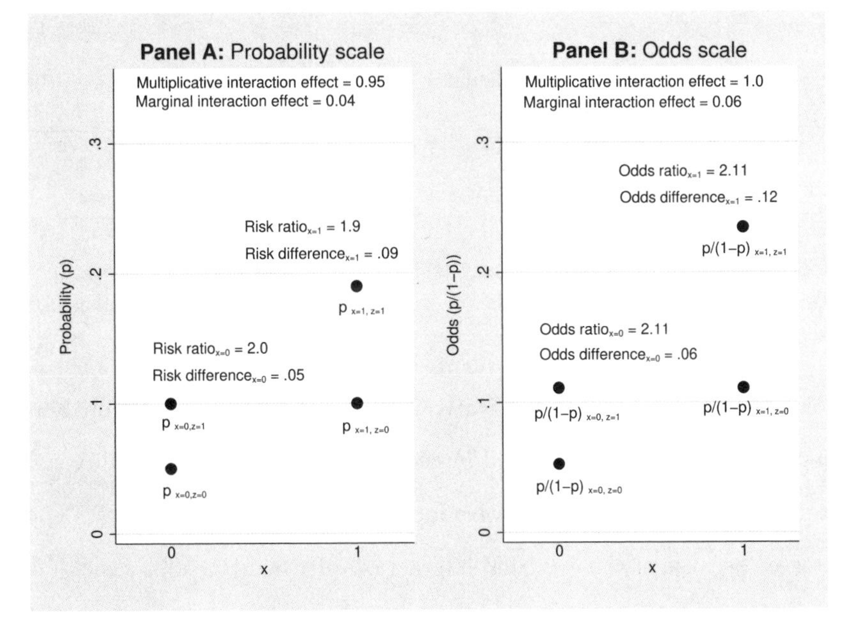

Figure 1. Multiplicative and marginal interaction effects

The multiplicative interaction effect on the probability scale is 0.95 (final column of table 1, panel A), which implies that as x increases, the effect of z decreases. By contrast, the multiplicative effect on the odds scale is 1.0 (final column of table 1, panel B), which implies that as x increases, the effect of z is unchanged. Finally, the marginal effect on the probability scale is 0.04 (middle column of table 1, panel A), which implies that as x increases, the effect of z increases. The marginal effect on the odds scale in panel B shows a similar increasing effect in this example, although in other examples it can differ meaningfully from the marginal effects on the probability scale.

Table 1. Main and interaction effects in nonlinear models

Panel A: Parameter calculations on the probability scale.

	Predicted probabilities (p)		Marginal effects (p)		Multiplicative effects (p)	
	$z=0$	$z=1$	Risk difference	Interaction effect	Risk ratio	Interaction effect
$x=0$	0.05	0.10	$0.10-0.05=0.05$	$0.09-0.05=0.04$	$\frac{0.10}{0.05}=2.0$	$\frac{1.9}{2.0}=0.95$
$x=1$	0.10	0.19	$0.19-0.10=0.09$		$\frac{0.19}{0.10}=1.9$	

Panel B: Parameter calculations on the odds scale.

	Predicted odds $(\frac{p}{1-p})$		Marginal effects $(\frac{p}{1-p})$		Multiplicative effects $(\frac{p}{1-p})$	
	z = 0	z = 1	Odds difference	Interaction effect	Odds ratio	Interaction effect
$x=0$	0.0526	0.1111	$0.1111-0.0526=0.06$	$0.12-0.06=0.06$	$\frac{0.1111}{0.0526}=2.1$	$\frac{2.1}{2.1}=1.0$
$x=1$	0.1111	0.2346	$0.2346-0.1111=0.12$		$\frac{0.2346}{0.1111}=2.1$	

Thus, in this simple example, the interaction effect could alternatively be interpreted as positive, null, or negative depending on which variant is estimated and reported. As x increases, the effect of z

- decreases because of the multiplicative interaction effect on the probability scale;
- increases because of the marginal interaction effect on the probability scale;
- is zero because of the multiplicative interaction effect on the odds scale; or
- increases because of the marginal interaction effect on the odds scale.

4 Discussion

To provide intuition regarding situations in which results are likely to differ across these variants, let's consider the case in which our example refers to a natural experiment. Suppose that z is an indicator for treatment ($z = 1$) versus control ($z = 0$) group and x is a time indicator of preperiod ($x = 0$) versus postperiod ($x = 1$). Our example is not well balanced in the preperiod: treatments had double the baseline risk as the controls. When there is such imbalance, it is well known that treatment-effect estimates are sensitive to functional form. In our example, the outcome in the control group doubled over time, an increase of 0.05; the outcome in the treatment group increased by a greater absolute amount of 0.09, but it did not quite double. Hence the different sign of the intervention effect: the marginal effect of 0.04 shows an increase in p in the treatment group relative to absolute growth in the controls, but the multiplicative interaction effect on the probability scale of 0.95 shows a negative effect of the treatment relative to the multiplicative growth in the controls. The preferred solution in the evaluation literature is to choose a different control group that is better matched at the baseline. With better matching, at least the signs (though not the magnitudes) of the interaction effect would be the same for the marginal and multiplicative effects.

Another important point is that our example, if it had included other covariates, would have different effect sizes across observations. In nonlinear models, the magnitudes of the marginal effects are not constant but vary across observations (Ai and Norton 2003). For example, although odds ratios are constant for all observations in a logistic model, marginal effects are typically larger when the underlying probability is close to 50% and smaller when the underlying probability is close to 0 or 1. On the other hand, the magnitude of the odds ratio from a logistic regression is scaled by an arbitrary factor that changes when additional covariates are added to the model, making comparisons of magnitudes impossible (Norton, Dowd, and Maciejewski 2018). However, marginal effects are more robust to changes in model specification. In summary, with a richer dataset, the researcher should be aware that treatment effects will often differ across observations. In this case, one can use the `margins` command to calculate average marginal effects—a summary measure of effect magnitudes and their statistical significance.

Different disciplinary traditions tend to default to different variants of these interaction effects. Setting aside the debates regarding the general merits and drawbacks of each (see Norton and Dowd [2018]; Mustillo, Landerman, and Land [2012]), we argue that it can be misleading to focus on only one variant by default. Thus, we build on Buis (2010) to argue that researchers should estimate both multiplicative and marginal interaction effects and report the sensitivity of key inferences.

References

Ai, C., and E. C. Norton. 2003. Interaction terms in logit and probit models. *Economics Letters* 80: 123–129. https://doi.org/10.1016/S0165-1765(03)00032-6.

Buis, M. 2010. Stata tip 87: Interpretation of interactions in nonlinear models. *Stata Journal* 10: 305–308. https://doi.org/10.1177/1536867X1001000211.

Doidge, C., G. A. Karolyi, and R. M. Stulz. 2013. The U.S. left behind? Financial globalization and the rise of IPOs outside the U.S. *Journal of Financial Economics* 110: 546–573. https://doi.org/10.1016/j.jfineco.2013.08.008.

Karaca-Mandic, P., E. C. Norton, and B. Dowd. 2012. Interaction terms in nonlinear models. *Health Services Research* 47: 255–274. https://doi.org/10.1111/j.1475-6773.2011.01314.x.

Kolasinski, A. C., and A. F. Siegel. 2010. On the economic meaning of interaction term coefficients in non-linear binary response regression models. https://papers.ssrn.com/sol3/papers.cfm?abstract_id=1668750.

Mustillo, S., L. R. Landerman, and K. C. Land. 2012. Modeling longitudinal count data: Testing for group differences in growth trajectories using average marginal effects. *Sociological Methods and Research* 41: 467–487. https://doi.org/10.1177/0049124112452397.

Norton, E. C., and B. E. Dowd. 2018. Log odds and the interpretation of logit models. *Health Services Research* 53: 859–878. https://doi.org/10.1111/1475-6773.12712.

Norton, E. C., B. E. Dowd, and M. L. Maciejewski. 2018. Odds ratios—Current best practice and use. *Journal of the American Medical Association* 320: 84–85. https://doi.org/10.1001/jama.2018.6971.

Vaidyanathan, B. 2011. Religious resources or differential returns? Early religious socialization and declining attendance in emerging adulthood. *Journal for the Scientific Study of Religion* 50: 366–387. https://doi.org/10.1111/j.1468-5906.2011.01573.x.

The Stata Journal (2020)
20, Number 1, pp. 244–249 DOI: 10.1177/1536867X20909707

Stata tip 135: Leaps and bounds

Maarten L. Buis
University of Konstanz
Konstanz, Germany
maarten.buis@uni-konstanz.de

A simple way of adding a variable nonlinearly to a model is to transform that variable. Common transformations are adding a quadratic term or taking a logarithm, but other transformations are also possible, such as taking the cube root (Cox 2011) or adding splines (see [R] **mkspline**). The purpose of this tip is to discuss yet another underused alternative transformation: the combination of continuous variables and indicator (dummy) variables.

Sometimes, a continuous variable consists of qualitatively different segments. A good example of such a variable is the number of hours a respondent usually works per week. In many countries, numbers less than 40 on such a variable represent respondents who work part-time, the number 40 represents respondents who work full-time, and numbers above 40 represent respondents who routinely work overtime. Using `nlsw88.dta`, which comes with Stata, we could analyze how one's number of hours worked per week influences one's average hourly wage, that is, the total earnings in a week divided by the hours worked that week. If we just add hours linearly, then we would conclude that an extra hour working is related to a four-cent increase in average hourly wage.

```
. sysuse nlsw88
(NLSW, 1988 extract)
. regress wage hours i.union i.race grade i.south, noheader
```

wage	Coef.	Std. Err.	t	P>|t|	[95% Conf.	Interval]
hours	.0425932	.0086062	4.95	0.000	.0257144	.059472
(output omitted)						

```
. quietly margins, at(union=0 race=1 grade=12 south=0) over(hours)
```

```
. marginsplot, noci plotopts(msymbol(i))
> ytitle("predicted hourly wage") title("")
  Variables that uniquely identify margins: hours
```

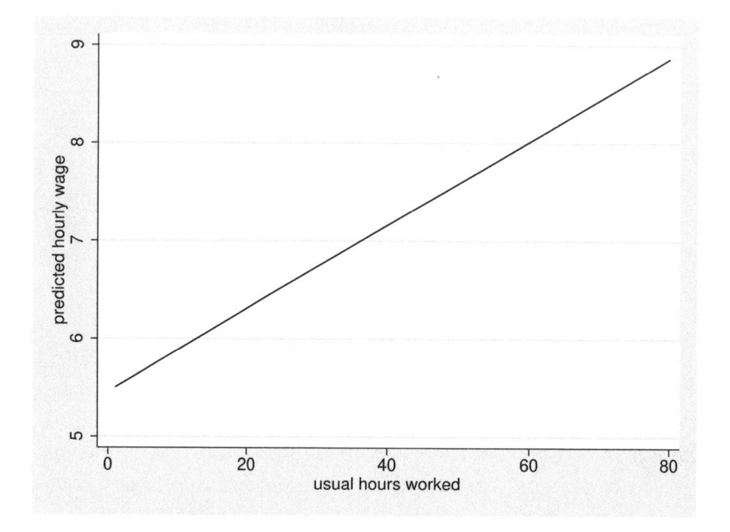

Figure 1. Linear effect of hours worked per week

However, we might hypothesize that working "normal" hours makes it easier for companies to standardize the allocation of tasks to the workers. As a consequence, companies might be willing to pay a premium for working full-time. This means that working more hours may increase average hourly wage, but there is an extra "jump" at 40. To test that, we can add both the variable **hours** and an additional indicator variable for full-time workers to our model. Cox and Schechter (2019) wrote a useful tutorial on how to effectively create indicator variables. In this model, an extra hour working is still associated with a 4-cent increase in average hourly wage, but those working full-time get a 35-cent "bonus". In this case, the indicator variable introduced a single spike at 40 hours worked per week.

```
. generate fulltime = hours == 40 if hours < .
(4 missing values generated)
. regress wage i.fulltime hours i.union i.race grade i.south, noheader
```

wage	Coef.	Std. Err.	t	P>|t|	[95% Conf.	Interval]
1.fulltime	.3486994	.1775229	1.96	0.050	.0005353	.6968636
hours	.0385415	.0088436	4.36	0.000	.0211972	.0558858
(output omitted)						

```
. quietly margins, at(union=0 race=1 grade=12 south=0) over(hours)
```

```
. marginsplot, noci plotopts(msymbol(i))
> ytitle("predicted hourly wage") title("")
  Variables that uniquely identify margins: hours
```

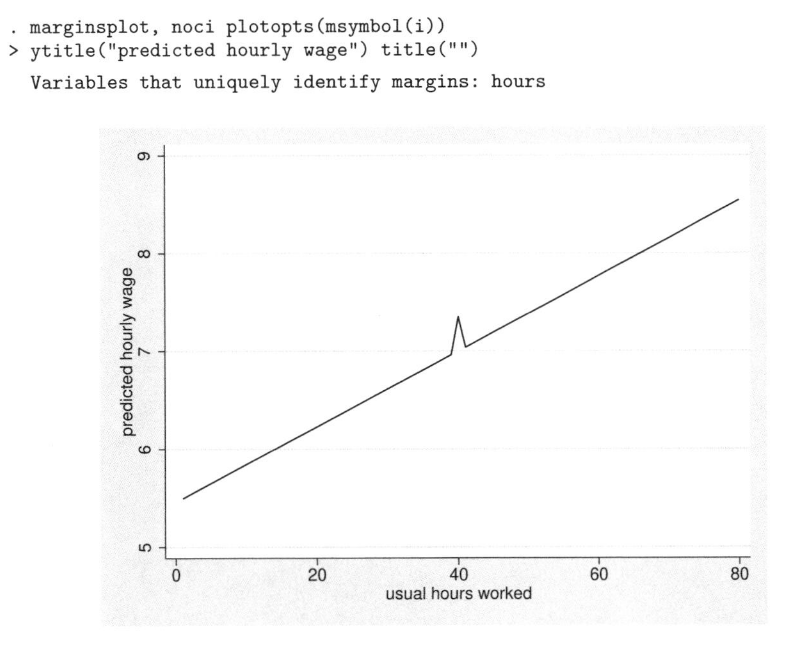

Figure 2. Linear effect of hours worked per week with a jump at working full-time

Sometimes, overtime is paid at a higher rate. So we might expect that working more hours generally increases the average hourly wage, but after 40 hours there is an extra jump that does not immediately disappear like before but persists. To test that, we can introduce the variable `hours` and an indicator variable for those respondents that routinely work overtime to the model. However, the results show that working overtime leads to a persistent (nonsignificant) 11-cent decrease in average hourly wage.

```
. generate overtime = hours > 40 if hours < .
(4 missing values generated)
. regress wage i.overtime hours i.union i.race grade i.south, noheader
```

wage	Coef.	Std. Err.	t	P>|t|	[95% Conf.	Interval]
1.overtime	-.1088363	.2742177	-0.40	0.691	-.6466419	.4289693
hours	.0449326	.0104327	4.31	0.000	.0244716	.0653936
(output omitted)						

```
. quietly margins, at(union=0 race=1 grade=12 south=0) over(hours)
```

```
. marginsplot, noci plotopts(msymbol(i))
> ytitle("predicted hourly wage") title("")
  Variables that uniquely identify margins: hours
```

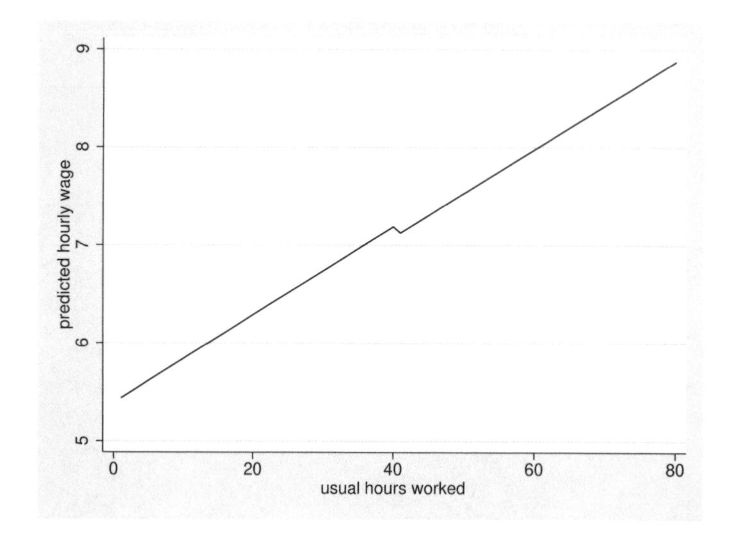

Figure 3. Linear effect of hours worked per week with a persistent jump for overtime

We forgot that not everybody gets his or her overtime paid. For those who get paid for working overtime, overtime will increase their average hourly wage. However, for those who are not paid for overtime, overtime will decrease their average hourly wage. We might expect that unpaid overtime happens in professions where people are intrinsically motivated (for example, academics), so they may work long hours. Whereas paid overtime happens in occupations where people are less intrinsically motivated, in which case both the workers and the employers have an incentive to keep the amount of overtime within bounds. So we hypothesize that the group of respondents working small amounts of overtime mainly consists of people getting paid overtime, while the group of respondents working large amounts of overtime consists mainly of people who do not get (completely) paid for overtime. In that case, we would expect a sharp increase in average hourly wage at 41 hours per week but a decrease after that. This is implemented by including an interaction between the overtime indicator variable and the `hours` variable. In this case, it makes sense to center the `hours` variable at 41; that way, the effect of the overtime indicator variable can be interpreted as the jump that occurs when one starts to work overtime. In this model, working an extra hour increases the average hourly wage by six cents if one works part time. If one starts working overtime, there is an immediate bonus of 1 dollar and 9 cents, but every extra hour decreases the average hourly wage by 11 cents ($6 - 17 = -11$). This type of regression is sometimes called segmented, broken-stick, or piecewise regression. This type of model is also closely related to a regression discontinuity design (Calonico, Cattaneo, and Titiunik 2014; Calonico et al. 2017).

```
. generate hours_c = hours - 41
(4 missing values generated)
. regress wage i.overtime##c.hours_c i.union i.race grade i.south, noheader
------------------------------------------------------------------------------
        wage |      Coef.   Std. Err.      t    P>|t|     [95% Conf. Interval]
-------------+----------------------------------------------------------------
  1.overtime |   1.090213   .3562397     3.06   0.002     .3915431    1.788883
     hours_c |   .0638511   .0109757     5.82   0.000     .0423253     .085377
             |
   overtime#|
   c.hours_c |
          1  |  -.1728014   .0331008    -5.22   0.000    -.2377198    -.107883
             |
  (output omitted)
------------------------------------------------------------------------------
. lincom 1.overtime#c.hours_c + hours_c
 ( 1)  hours_c + 1.overtime#c.hours_c = 0
------------------------------------------------------------------------------
        wage |      Coef.   Std. Err.      t    P>|t|     [95% Conf. Interval]
-------------+----------------------------------------------------------------
         (1) |  -.1089503   .0312445    -3.49   0.000    -.1702281   -.0476725
------------------------------------------------------------------------------
. quietly margins, at(union=0 race=1 grade=12 south=0) over(hours)
. marginsplot, noci plotopts(msymbol(i))
> ytitle("predicted hourly wage") title("")
  Variables that uniquely identify margins: hours
```

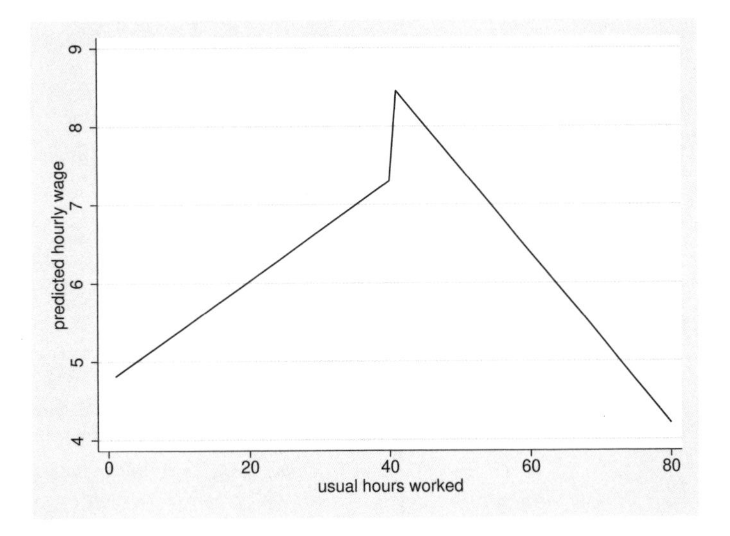

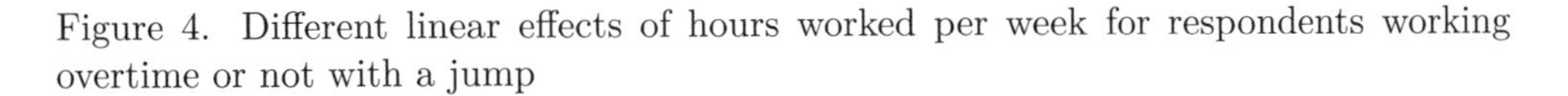
Figure 4. Different linear effects of hours worked per week for respondents working overtime or not with a jump

By combining continuous and indicator variables, one can allow for nonlinearity by adding spikes, persistent jumps, or complete breaks to the regression line. This flexibility allows one to tailor the kind of nonlinearity in the model to the research question and what one knows about the variables involved with only a few parameters. Moreover, those parameters are easy to interpret.

References

Calonico, S., M. D. Cattaneo, M. H. Farrell, and R. Titiunik. 2017. rdrobust: Software for regression-discontinuity designs. *Stata Journal* 17: 372–404. https://doi.org/10.1177/1536867X1701700208.

Calonico, S., M. D. Cattaneo, and R. Titiunik. 2014. Robust data-driven inference in the regression-discontinuity design. *Stata Journal* 14: 909–946. https://doi.org/10.1177/1536867X1401400413.

Cox, N. J. 2011. Stata tip 96: Cube roots. *Stata Journal* 11: 149–154. https://doi.org/10.1177/1536867X1101100112.

Cox, N. J., and C. B. Schechter. 2019. Speaking Stata: How best to generate indicator or dummy variables. *Stata Journal* 19: 246–259. https://doi.org/10.1177/1536867X19830921.

The Stata Journal (2020)
20, Number 2, pp. 489–492 DOI: 10.1177/1536867X20931008

Stata tip 136: Between-group comparisons in a scatterplot with weighted markers

Andrew Musau
Inland School of Business & Social Sciences
INN University
Lillehammer, Norway
andrew.musau@inn.no

Scatterplots are a convenient tool to represent the relationship between two continuous variables. Often, it is necessary to classify this relationship according to values of a third categorical variable. There are several ways to do this (see, for example, Cox [2005]). Here we consider a variation of these graphs, sometimes referred to as bubbleplots, where an additional dimension of the data is represented in the size of the markers. In Stata, one can create such a graph by explicitly specifying a weight in the standard scatterplot syntax (see [G-2] **graph twoway scatter**). As an example, we will use `auto.dta`. Suppose we want to create a scatterplot of mileage and weight with markers weighted by repair record. Suppose further that we want to compare domestic (American) and foreign cars. A problem arises if we do not observe all values of the weighting variable in each of the groups defined by the categorical variable. Consider a cross-tabulation of repair record and car type.[1]

```
. sysuse auto
(1978 Automobile Data)

. tabulate rep78 foreign
```

Repair Record 1978	Car type Domestic	Foreign	Total
1	2	0	2
2	8	0	8
3	27	3	30
4	9	9	18
5	2	9	11
Total	48	21	69

All five values of repair record are observed for domestic cars, whereas two values are not present for foreign cars. A comparison of scatterplots of mileage and weight with markers weighted by repair record for all cars in the dataset and for groups defined by car type results in the pair of graphs shown in figure 1.

```
. twoway (scatter mpg weight [aweight = rep78], mcolor(black)
> msymbol(smcircle_hollow) text(31 2400 "2", color(black))
> text(41 2240 "1", color(black)) text(18 2550 "4", color(black))
> text(21 2330 "3", color(black)) scheme(sj) legend(on order(1 "All cars")))
```

1. I thank an anonymous referee for suggesting improvements to the presentation of the problem and proposed solutions.

```
. twoway (scatter mpg weight [aweight = rep78] if foreign==0, mcolor(gs5)
> msymbol(smcircle_hollow))(scatter mpg weight [aweight = rep78] if foreign==1,
> mcolor(gs11) msymbol(smcircle_hollow)
> legend(order(1 "American" 2 "Foreign") row(1))
> text(31 2350 "2", color(black)) text(41 2190 "1", color(black))
> text(18 2560 "4", color(black)) scheme(sj) text(21 2280 "3", color(black)))
```

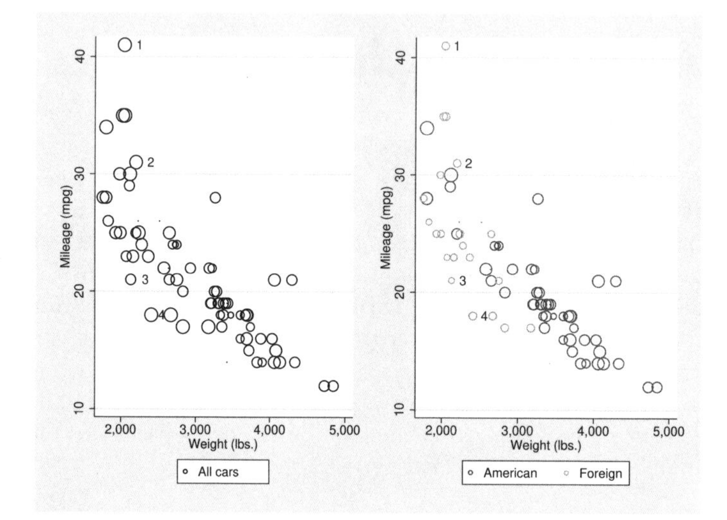

Figure 1. Scatterplot with all values of the weighting variable not present in each group

The size of the weighted markers corresponding to foreign cars is smaller on the graph on the right-hand side, as can be seen from the selection of markers numbered in figure 1. The issue is that Stata internally rescales the weights within groups, thereby precluding between-group comparisons. Note that the problem also arises if the graphs are created with the commonly used `by()` option.

```
. twoway scatter mpg weight [aweight = rep78], by(foreign, total)
> mcolor(gs5 gs11) msymbol(smcircle_hollow smcircle_hollow)
> legend(order(1 "American" 2 "Foreign") row(1))
> text(31 2350 "2", color(black)) text(41 2190 "1", color(black))
> text(18 2560 "4", color(black)) text(21 2280 "3", color(black))
> scheme(sj)
```

The top panel of figure 2 resembles figure 1, while the bottom panel illustrates the two proposed solutions. The first solution is to add "pseudo-observations" to the dataset to ensure that all values of the weighting variable are present in each group of the categorical variable. However, an ensuing concern is that these extra observations will distort the resulting graph. Fortunately, this is not the case if the added observations are missing values for the continuous variables. The command `fillin` (see [D] **fillin**) allows us to achieve this by adding observations with missing data so that all interactions of car type and repair record exist.

```
. fillin foreign rep78
```

A cross-tabulation of repair record and car type will now confirm that each group of the latter includes all values of the former. The second proposed solution, drawing on Cox (2005), is to use the command `separate` (see [D] **separate**). The underlying mechanics between both approaches are basically the same. For each variable, `separate` produces missing values in the continuous variable in all but one group, yet the weights are rescaled based on all observations.

```
. separate mpg, by(foreign)
  (output omitted)
. twoway scatter mpg? weight [aweight = rep78], mcolor(gs5 gs11)
> msymbol(smcircle_hollow smcircle_hollow)
> legend(order(1 "American" 2 "Foreign") row(1))
> text(31 2350 "2", color(black)) text(41 2190 "1", color(black))
> text(18 2560 "4", color(black)) text(21 2280 "3", color(black))
> scheme(sj)
```

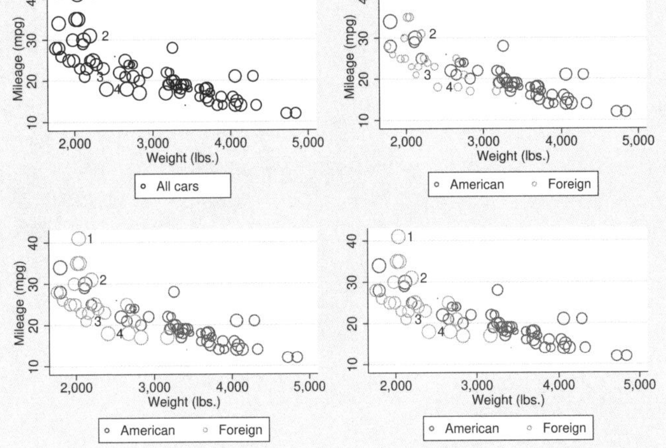

Figure 2. Scatterplots before and after implementing suggested solutions

In summary, the `fillin` approach adds observations to the dataset corresponding to the number of nonexistent values of the weighting variable across groups defined by the categorical variable. On the other hand, the `separate` approach adds variables to the dataset corresponding to the number of groups in the categorical variable. While extra observations created by `fillin` do not distort the graph, they can distort other analyses (for example, as illustrated by the proposed cross-tabulation) and should be deleted once the graphs are created. These are marked by the `_fillin` variable, which should be used to revert to the original dataset. For `separate`, the added variables may subsequently be deleted, but their presence in the dataset does no harm. If you use the `if` qualifier in the `graph twoway` command, you should use the `separate` approach because the syntax is shorter. On the other hand, if you are creating graphs by groups defined by the categorical variable, you should use the `fillin` approach because you can use the `by()` option. To achieve the same result using the `separate` approach, you would need to create one graph at a time and thereafter use the command `graph combine` (see [G-2] **graph combine**) to merge these graphs. Finally, in terms of efficiency, because of maximum size limits, it would appear that adding variables to the dataset is costlier than adding observations.[2] However, because only a few groups can reasonably be differentiated in the scatterplot with weighted markers, comparing the approaches practically, this difference matters little. Therefore, preference for one over the other is a matter of taste.

Reference

Cox, N. J. 2005. Stata tip 27: Classifying data points on scatter plots. *Stata Journal* 5: 604–606. https://doi.org/10.1177/1536867X0500500412.

2. In Stata/SE 16, for example, the current maximum limit for observations is approximately 2,147 million compared with only 32,767 for variables.

The Stata Journal (2020)
20, Number 2, pp. 493–498 DOI: 10.1177/1536867X20931027

Stata tip 137: Interpreting constraints on slopes of rank-deficient design matrices

Demetris Christodoulou
University of Sydney
Sydney, Australia
demetris.christodoulou@sydney.edu.au

A rank-deficient design matrix of explanatory variables $\mathbf{X}$ is not of full-column rank when there is one or more linear dependencies, meaning that $\mathbf{X}'\mathbf{X}$ is singular and its inverse does not exist; thus, there is no unique solution to $\mathbf{b} = (\mathbf{X}'\mathbf{X})^{-1}\mathbf{X}'\mathbf{y}$. Rank deficiency is sometimes referred to as "perfect collinearity".

There are two ways to enable the use of $\mathbf{X}$ in regression analysis. If there is one linear dependency, then the standard approach is to reduce the dimension of $\mathbf{X}$ by identifying a zero-parameter constraint on one of its columns. This is the default treatment in Stata and other software, that is, to arbitrarily remove one column of $\mathbf{X}$.

The alternative approach is to expand $\mathbf{X}$ by adding an extra column through the identification of a linear constraint across the parameters. Then, $\mathbf{X}$ can be used in constrained least squares via the Stata command `cnsreg`. Both approaches yield the same fully identified model, but the interpretation of their estimated coefficients depends entirely on the imposed constraint.

Much of the relevant econometric literature focuses on the identification of rank-deficient matrices of mutually exclusive binary variables and the interpretation of their intercepts, a problem so ubiquitous and well understood that it has earned its colloquial moniker of "dummy variable trap". The interpretation of constrained intercepts is indeed elementary because it is a simple matter of weighted constants.

However, as I discuss in Christodoulou (2018), a more rigorous discussion on the effect on slope coefficients of a rank-deficient $\mathbf{X}$ seems to evade the literature. When $\mathbf{b}$ involves slope coefficients, the reduction of $\mathbf{X}$ by imposing zero-parameter constraints or the expansion of $\mathbf{X}$ by imposing linear constraints amounts to an imposition of a structural relation on the parameters to be estimated. The interpretation of the constrained slopes then becomes conditional on the validity of the structural constraint.

Consider the question of how capital investment in operating assets affects sales revenue in fixed asset-intensive firms. Companies with high stakes in tangible assets rely on capital investment to boost revenue, but the more the assets are used in operations, the more their value is depleted and needs to be replenished. The economic transactions describing this relation are captured by the accounting identity

$$\text{ppe}_{it-1} + \text{cpx}_{it} - \text{dep}_{it} \equiv \text{ppe}_{it} \tag{1}$$

or equivalently stated as $\Delta\text{ppe}_{it} \equiv \text{cpx}_{it+1} - \text{dep}_{it+1}$, where ppe_t is the stock in property plant and equipment, cpx_{t+1} is new capital expenditure, and dep_{t+1} is depreciation plus other events that may deplete assets, such as the sale of assets. Naturally, increases in capital stock are expected to bring more sales. Indeed, one could argue that the variation in the period's sales could be explained by the average capital investment used from t to $t+1$.

The following simple simulation generates data that describe this scenario:

```
. set type double
. set seed 1234
. set obs 10000
number of observations (_N) was 0, now 10,000
. generate id = _n
. generate ppe0 = rnormal(4.5,1)
. generate cpx1 = ppe0*0.08 + rnormal(0.5,0.2)
. generate dep1 = (ppe0+cpx1)*0.06 + rnormal(0.3,0.1)
. generate ppe1 = ppe0 + cpx1 - dep1
. generate sales = 0.25 + ((ppe0+ppe1)/2)*0.1 + rnormal(0,0.1)
```

The parameters of the random normal distributions are selected so that they appear somewhat realistic, considering that these are a result of log-transformations from originally log-normally distributed variables.

Let's say that someone is interested in learning how much revenue would change if a company decides to spend more in new capital expenditure and, at the same time, how depletion would affect sales, conditional of course on the capital investment stock. Then, the variation of sales revenue could be written as a function of the structural relation of (1) plus a random-error term:

$$\text{sales}_{it} = a + b_1\text{ppe}_{it-1} + b_2\text{cpx}_{it} + b_3\text{dep}_{it} + b_4\text{ppe}_{it} + \epsilon_{it} \tag{2}$$

Given the rank deficiency in $\mathbf{X}$, Stata will estimate this regression by imposing a zero-parameter restriction to one of the explanatory variables and also issue a warning that a variable was omitted because of perfect collinearity:

```
. regress sales ppe0 cpx1 dep1 ppe1, noheader
note: dep1 omitted because of collinearity
```

sales	Coef.	Std. Err.	t	P>|t|	[95% Conf.	Interval]
ppe0	.0500818	.0095262	5.26	0.000	.0314086	.068755
cpx1	.0073934	.0106667	0.69	0.488	-.0135154	.0283023
dep1	0	(omitted)				
ppe1	.0470199	.0100562	4.68	0.000	.0273077	.0667321
_cons	.2559735	.0060155	42.55	0.000	.2441819	.2677651

Stata decided to drop the variable `dep1`, but this could have been another explanatory variable; for example, changing the seed to `1235` would drop `cpx1`. The interpretation of the remaining estimated slope parameters depends on the validity of the zero-parameter restriction on $b_3 = 0$ on `dep1`, a highly doubtful assumption even with real data. Let's see what happens when we estimate all competing specifications with zero-parameter constraints, that is, each time omitting one explanatory variable:

```
. quietly regress sales cpx1 dep1 ppe1
. estimates store ppe0_0
. quietly regress sales ppe0 dep1 ppe1
. estimates store cpx1_0
. quietly regress sales ppe0 cpx1 ppe1
. estimates store dep1_0
. quietly regress sales ppe0 cpx1 dep1
. estimates store ppe1_0
. estimates table ppe0_0 cpx1_0 dep1_0 ppe1_0, se(%5.4f) stats(rmse ll)
```

Variable	ppe0_0	cpx1_0	dep1_0	ppe1_0
cpx1	-.04268835		.00739343	.05441336
	0.0055		0.0107	0.0051
dep1	.05008178	.00739343		-.04701994
	0.0095	0.0107		0.0101
ppe1	.09710171	.05441336	.04701994	
	0.0012	0.0051	0.0101	
ppe0		.04268835	.05008178	.09710171
		0.0055	0.0095	0.0012
_cons	.25597348	.25597348	.25597348	.25597348
	0.0060	0.0060	0.0060	0.0060
rmse	.09990187	.09990187	.09990187	.09990187
ll	8848.2841	8848.2841	8848.2841	8848.2841

legend: b/se

Note how the magnitudes of the estimated slopes switch place depending on the variable that is omitted from estimation. This is because each restriction sways estimation so that the collection of all estimated slopes remains parallel to the null vector that describes the linear dependency (for an illustration, see figure 1 in Christodoulou [2018]). This sort of behavior makes any discussion on marginal effects entirely meaningless.

Such ad hoc imposed constraints, whose only purpose is to enable mere estimation, are dangerous practices when applied on rank-deficient design matrices involving slope coefficients. A zero-parameter restriction on a slope suggests a zero marginal effect, and in this case such restrictions are simply untenable.

Another way to enable estimation is to expand $\mathbf{X}$ by imposing a linear constraint that specifies a structural relation across all parameters. For example, one could suggest that (2) behaves like a homogeneous function of some degree. The structure of the data does not allow us to estimate which degree this is, so we need to assume the degree as a constraint. We could claim that the slope coefficients add to some fixed c, thus effectively imposing a homogeneous function of degree c, meaning that a fixed change in all independent variables would change the dependent variable by that value raised to the power of c.

For example, for $c = 0$, a fixed change would result in no change in the dependent variable. For $c = 1$, a fixed change would result in a linear change in the dependent variable, or what the economists call a "constant return to scale" within the right context; for $c < 1$, we have decreasing returns to scale, and for $c > 1$ we have increasing returns to scale. Consider the following examples:

```
. constraint define 1 ppe0 + cpx1 - dep1 - ppe1 = 0
. quietly cnsreg sales ppe0 cpx1 dep1 ppe1, collinear constraint(1)
. estimates store c0
. constraint define 1 ppe0 + cpx1 - dep1 - ppe1 = 0.75
. quietly cnsreg sales ppe0 cpx1 dep1 ppe1, collinear constraint(1)
. estimates store c0p75
. constraint define 1 ppe0 + cpx1 - dep1 - ppe1 = 1
. quietly cnsreg sales ppe0 cpx1 dep1 ppe1, collinear constraint(1)
. estimates store c1
. constraint define 1 ppe0 + cpx1 - dep1 - ppe1 = 1.25
. quietly cnsreg sales ppe0 cpx1 dep1 ppe1, collinear constraint(1)
. estimates store c1p25
. estimates table c0 c0p75 c1 c1p25, se(%5.4f) stats(rmse ll)
```

Variable	c0	c0p75	c1	c1p25
ppe0	.04746796	.23496796	.29746796	.35996796
	0.0028	0.0028	0.0028	0.0028
cpx1	.00477961	.19227961	.25477961	.31727961
	0.0045	0.0045	0.0045	0.0045
dep1	.00261382	-.18488618	-.24738618	-.30988618
	0.0073	0.0073	0.0073	0.0073
ppe1	.04963375	-.13786625	-.20036625	-.26286625
	0.0031	0.0031	0.0031	0.0031
_cons	.25597348	.25597348	.25597348	.25597348
	0.0060	0.0060	0.0060	0.0060
rmse	.09990187	.09990187	.09990187	.09990187
ll	8848.2841	8848.2841	8848.2841	8848.2841

legend: b/se

The option `collinear` tells Stata to keep perfectly collinear variables, thus ensuring reporting of all estimated coefficients.

Note how the addition of all estimated slopes is always the same, at $\widehat{b}_1 + \widehat{b}_2 + \widehat{b}_3 + \widehat{b}_4 = 0.10449514$, regardless of the imposed constraint. This is the same constant to the addition of the coefficients as with the estimates with zero-parameter constraints, as above. Regardless of the constraint, the coefficients must add up to the same constant.

The coefficients are simply scaled up or down by a fixed amount as c changes. This means that because the constraints are needed for identification, the rank-deficient nature of the data does not allow one to say which structural constraint is most appropriate. One must assume it.

The model with the homogeneous function of degree zero, with $c = 0$, can also be fit using the Moore–Penrose pseudoinverse (for example, see Mazumdar, Li, and Bryce [1980]; Searle [1984]), using the `pinv()` Mata function as follows:

```
. mata:
------------------------------------------------- mata (type end to exit) ------
: y = st_data(.,("sales"))
: X = st_data(., ("ppe0" ,"cpx1", "dep1", "ppe1"))
: n = rows(X)
: X = X,J(n,1,1)
: XpXi = pinv(quadcross(X,X))
: b = XpXi*quadcross(X,y)
: end
--------------------------------------------------------------------------------
. mata: transposeonly(b)
                 1              2              3              4              5
    +-----------------------------------------------------------------------------+
  1 |  .0474679606    .0047796107    .0026138174    .0496337539    .2559734849  |
    +-----------------------------------------------------------------------------+
```

These are identical coefficients to those reported in the table just above under the heading c0, in that order. Using the Moore–Penrose pseudoinverse, we can recover every other solution that is parallel to the null vector. Given that there are four coefficients, $k = 4$, then the imposition of an assumed degree for the homogeneous function c must be equally allocated across the k coefficients. For instance, for $k = 0.75$, it holds that

```
. mata: b[1] + 0.75/4, b[2] + 0.75/4, b[3] - 0.75/4, b[4] - 0.75/4
                  1                2                3                4
    +-------------------------------------------------------------------+
  1 |   .2349679606     .1922796107    -.1848861826    -.1378662461    |
    +-------------------------------------------------------------------+
```

and similarly for any other c. Similarly, because the Moore–Penrose pseudoinverse gives the solution for $c = 0$, we could use this result to see what would be the set of estimates for any given zero-parameter restriction. Here is the case of the zero-parameter restriction on the coefficient of ppe_{it-1}, which is the same as that reported in the first column of the first estimates table above:

```
. mata: b[1] - b[1], b[2] - b[1], b[3] + b[1], b[4] + b[1]
                  1                2                3                4
    +-------------------------------------------------------------------+
  1 |             0    -.0426883499      .050081778     .0971017145    |
    +-------------------------------------------------------------------+
```

Finally, an important note about standard errors—they remain the same across all specifications. As shown in Greene and Seaks (1991), the individual standard errors of regressions involving rank-deficient design matrices are no longer informative. We cannot speak of coefficient-specific statistical significance. For example, in the first table of estimates reported, the coefficient on cpx1 appears as statistically insignificant with a p-value of 0.488. This standard-error estimate is of course nonsensical, given the nature of the simulated data. In specifications with rank-deficient design matrices, we can speak only about the fit of the overall model, as in the root mean squared error and the estimated log likelihood, which remain identical regardless of the type of constraint. There is no such thing as coefficient significance.

In Christodoulou and McLeay (2014, 2019), we use Stata to explain how this lack of insight has proven to be an acute problem in financial research that relies on inputs from the rank-deficient accounting data matrix of articulated financial statements. Accounting data, governed by a double-entry data-generating process whereby a transaction is recorded twice, is purposefully designed to be rank deficient of order one. This is a matter of structural nonidentification and requires the additional specification of a suitable constraint to enable estimation. If the constraint is arbitrarily imposed, then inference is entirely useless.

Acknowledgment

I acknowledge the useful comments by an anonymous reviewer.

References

Christodoulou, D. 2018. The accounting identity trap: Identification under stock-and-flow rank deficiency. *Applied Economics* 50: 1413–1427. https://doi.org/10.1080/00036846.2017.1363860.

Christodoulou, D., and S. McLeay. 2014. The double entry constraint, structural modeling and econometric estimation. *Contemporary Accounting Research* 31: 609–628. https://doi.org/10.1111/1911-3846.12038.

———. 2019. The double entry structural constraint on the econometric estimation of accounting variables. *European Journal of Finance* 25: 1919–1935. https://doi.org/10.1080/1351847X.2019.1667847.

Greene, W. H., and T. G. Seaks. 1991. The restricted least squares estimator: A pedagogical note. *Review of Economics and Statistics* 73: 563–567. https://doi.org/10.2307/2109587.

Mazumdar, S., C. C. Li, and G. R. Bryce. 1980. Correspondence between a linear restriction and a generalized inverse in linear model analysis. *American Statistician* 34: 103–105. https://doi.org/10.1080/00031305.1980.10483009.

Searle, S. R. 1984. Restrictions and generalized inverses in linear models. *American Statistician* 38: 53–54. https://doi.org/10.1080/00031305.1984.10482873.

The Stata Journal (2020)
20, Number 2, pp. 499–503 DOI: 10.1177/1536867X20931028

Stata tip 138: Local macros have local scope

Nicholas J. Cox
Department of Geography
Durham University
Durham, UK
n.j.cox@durham.ac.uk

1 Introduction

Stata programs, usually, and other Stata code chunks, commonly, use local macros. This tip focuses on a common misunderstanding of how local macros work. The executive summary is given in the title.

The best introduction to local macros remains material in [U] **18 Programming Stata**, but the tip is written to be self-contained as far as possible.

2 What is a macro?

The term *macro* has various meanings, even within programming, let alone beyond. In some software, it has informal or even formal meaning as indicating a routine or script, typically containing a series of instructions or statements in that software. Be that as it may, in Stata a macro is a named container whose contents are text. A macro is just one item, although occasionally that item may be very large.

Stata's use of macros is not unusual, either. It is akin in spirit to other languages, particularly languages like C or those in the Unix tradition (Kernighan and Pike 1984; Kernighan and Ritchie 1988).

Macros in Stata have two flavors, global and local. Part of the misunderstanding discussed here is not realizing that those names are not just arbitrary jargon but explain the scope of the macro, namely, how widely it will be understood. We will get to the nub of the problem, slowly but surely, by looking at global macros first.

3 Global macros—and their limitations

Suppose you are working with Stata's auto data,

```
. sysuse auto
(1978 Automobile Data)
```

and you want to think of those variables that measure various size characteristics as one group. You can put their names into a global macro with

```
. global size "headroom trunk weight length displacement"
```

The text within " " has been copied as contents to the global macro just created with the name `size`. That allows you to type `$size` anywhere in your code knowing that Stata will use the definition of the macro to replace the macro name, `size`, with its contents, a string including various variable names. So you might be interested in a regression predicting car price as a response, contained in the variable `price`. Typing

```
. regress price $size
  (output omitted)
```

would save you typing out those variable names again or even selecting them individually otherwise.

The key here: When Stata sees that command, the reference to the global macro `size` will be evaluated, meaning that the macro name will be replaced by its contents so that the rest of Stata will see

```
. regress price headroom trunk weight length displacement
  (output omitted)
```

Note the syntax detail: the dollar sign `$` flags that a global macro name follows.

Here is another example. Suppose we want to obtain and hold the median price for some purpose, say, to split prices into those above and below the median or to use it in preparing a graph. For that, we could do

```
. summarize price, detail
  (output omitted)
. global median_price = r(p50)
. display "$median_price"
5006.5
```

If you are thinking that `5006.5` looks like a number, you are naturally right, but that is not a problem or an exception. The definition that macros hold text still holds: text being entirely or partially composed of numeric characters is fine. The rest of Stata will make its own judgments according to what it sees once the macro contents have been substituted.

A little analogy that may help is to think that a macro is a kind of bag or box in which text is stored, just as you might put a pen or some other possession in a bag or box for safekeeping. Once you have taken the text out of the bag or box, its ever having been inside has no implications for how it is used.

Where might a global macro be used? Within one or more of the following:

1. an interactive session in which commands are typed in the Command window,
2. a do-file you run,
3. code you run from a Do-file Editor, or
4. a program you write or use.

This may sound good, but it is not as good as it may sound. The sales pitch is that you can make one convenient definition of a global macro, and it will be understood everywhere you use that global macro. But that universal visibility might go horribly wrong. Commands can call other commands, do-files can call commands, and do-files can call other do-files. And yet other things can happen in code. In any of these ways, if you define a global macro, you unwittingly may undermine some other code that uses the same global macro name. Stata does not know or care who wrote that code, or when, and does not discriminate in any way. If you ask for global scope, then global scope is what you get.

You may think of work-arounds. One is just to use macro names so bizarre that in practice no one else can possibly have used them before, including yourself in some previous state of existence. But I do not want to have to use bizarre names. I guess that you really do not want that obligation either. Concise but evocative names are always valuable, and you do not want to have to ensure that a name differs from any already in use.

4 Local macros

So, to cut to the chase, the answer is to use local macros. Local macros have local scope (only). Shortly, we will spell out precisely what that means. But the aim is to be always safe, and never sorry. You are completely in charge of how that local macro is used. You are guaranteed that fears and worries about whether you used that name somewhere else—or more crucially whether someone else did in his or her code that you are using—are utterly groundless.

Otherwise, local macros are just like global macros. They contain text. That text may include numeric characters. What is crucial is what the rest of Stata does when the contents of local macros are revealed by evaluation.

So, if I wrote interactively

```
local size "headroom trunk weight length displacement"
```

then now I can refer to that local macro interactively in the same session, just as I could refer to a global macro. I could issue a regression command,

```
regress price `size´
```

or any other appropriate command,

```
summarize `size´
pca `size´
```

I can also use a definition given in a do-file later in that do-file.

There is always a downside. Here is where being "local" bites. Local means local scope, so in particular:

- A local macro defined in an interactive session is not visible in a do-file or to the code you are running from a Do-file Editor (which may he held in some kind of temporary file). And all of those limits apply in reverse. More generally, each of those is a separate place and a local macro defined in one is not visible in any of the others.
- Perhaps surprisingly: If you choose to run just chunks of code (which may be as short as individual lines of code) selected from a larger section in a Do-file Editor, "local" now means within the same chunk of code. It is not enough that a local macro was defined earlier in the window. The definition must be included in the chunk being run. When this bites, caution on the part of the user is not being rewarded. You may be feeling your way through some code, perhaps checking what happens at each step in a long process, or some of your code may have been revised and part of the analysis is being repeated in slightly different form. Style of working, however, has no implications. The question is only whether the (last) macro definition, the `local` command defining the macro, is visible within the same code chunk.

I close with two further comments.

What happens when the definition of a local macro is not visible?

It is as if that local macro did not exist. Hence, references to that local macro are replaced with empty strings. Otherwise put, the reference is just blanked out. The consequence is not necessarily illegal code. The consequence is unlikely, however, to be what you want. Suppose, for example, that the definition of local macro `size` were not visible, because it is not within the chunk of code being run. In that circumstance,

```
regress price `size'
```

is just interpreted as

```
regress price
```

That is legal: the instruction is to fit a regression to `price` alone. The result is just a regression fit in which the mean of price will be returned as intercept. No predictors will appear in the model because none were supplied. As another example,

```
summarize `size'
```

is just interpreted as

```
summarize
```

if the definition of the local macro is not visible. Again, that is legal: it means `summarize` all variables, which will work so long as there are some variables to summarize.

An exercise if you want one: what would happen if Stata saw only the bare command `pca`?

Referring to a local macro that does not exist is not itself a syntax error. The consequence may be, as above, something legal, which is not what you want. Or it may be something that is illegal for different reasons. Such subtleties can mean that a bug is hard to find, but being aware of this pitfall is a good start.

Why is the division between local and global so sharp? Are there not intermediate cases?

Indeed, sometimes a Stata user wants a combination of local and global approaches for good reasons, typically that a bundle of definitions makes sense for all the code files for a particular project. The `include` command is designed for these situations. For more detail, see [P] **include** or Herrin (2009).

References

Herrin, J. 2009. Stata tip 77: (Re)using macros in multiple do-files. *Stata Journal* 9: 497–498. https://doi.org/10.1177/1536867X0900900310.

Kernighan, B. W., and R. Pike. 1984. *The UNIX Programming Environment*. Englewood Cliffs, NJ: Prentice Hall.

Kernighan, B. W., and D. M. Ritchie. 1988. *The C Programming Language*. 2nd ed. Englewood Cliffs, NJ: Prentice Hall.

 DOI: 10.1177/1536867X20976341

The Stata Journal (2020)
20, Number 4, pp. 1016–1027

Stata tip 139: The by() option of graph can work better than graph combine

Nicholas J. Cox
Department of Geography
Durham University
Durham, UK
n.j.cox@durham.ac.uk

1 by() option of graph or graph combine for paneled figures?

Stata users produce graphs (often called figures) for their presentations and publications. Many such figures are composites with two or more separate panels, so that figure 1 is a composite of figure 1a, 1b, 1c, and 1d. There are two main ways to create such composites in Stata: using a `by()` option or using `graph combine` on previously created graphs. There are also commands such as `graph matrix` whose purpose is to create plots with multiple panels directly.

A `by()` option is likely to seem obvious when a grouping variable is already defined, as when you go something like

```
. sysuse auto
. scatter mpg weight, by(foreign)
```

to get separate scatterplots of `mpg` (miles per gallon) versus `weight` for domestic and foreign cars, domestic cars being those made in the United States and foreign cars being those made outside. The result of those commands may already be familiar to you. If not, running the commands just given is an excellent simple exercise.

Using `graph combine` is most obvious when graphs created separately nevertheless have point in being presented together. There might be some pedagogic or even polemical purpose to comparing a histogram, a box plot, a quantile plot, and a kernel density estimate as ways of showing a univariate distribution, each with advantages and disadvantages. It is worth flagging that paneling that is 2×2, 3×3, and so on can look good, although problems with odd numbers of panels such as 3, 5, or 7 will occur with their own obstinate frequency.

The aim of this tip is to push a little at this distinction, as less clear-cut than it seems. Sometimes, a graph will be improved by seeking a different data layout in which a `by()` option ensures more uniformity and consistency of style than will result from using `graph combine`.

2 Example 1: Confidence intervals for means of different variables

A standard plot in many fields compares sample means of a variable for two or more conditions within a framework of confidence intervals or other indications of uncertainty about those means. The word *confidence* here is, as every good text or course should explain, a term of art, rather than an expression of the conviction or intensity of belief a researcher is entitled or expected to hold about variability. A personal suspicion that *diffidence intervals* would be a better term is far from original. Almost a century ago, Fisher (1925, 10) wrote of "our mental confidence or diffidence" in making inferences, although he was not writing about confidence intervals *avant la lettre*; he was flagging the concept of likelihood.

A natural extension when there are several variables of interest is to show various such plots side by side.

An excellent way to get such plots is through `statsby` and `ci` (Cox 2010). The reduction to a new dataset with means and bounds for confidence intervals is immediate. There is some inefficiency in doing that again and again, as we will do here, but coding otherwise is likely to be more time consuming. The default confidence level with `ci` and more generally is 95%.

Let us suppose that—using the auto data again, and now for real—we want to show, side by side, plots with means and confidence intervals for price, miles per gallon, weight, and length for those domestic and foreign cars.

The plan here is to just loop over four variables, each time reading in the data, producing a reduced dataset and drawing and saving each graph for later combination. For an introduction to loops, and their use of local macros, see Cox (2020).

```
. tokenize "price mpg weight length"
. set scheme sj
. forvalues k = 1/4 {
  2. sysuse auto, clear
  3. if "``k''" == "price" label var price "Price (USD)"
  4. local which : variable label ``k''
  5. if "`which'" == "" local which "``k''"
  6. statsby, by(foreign) clear: ci mean ``k''
  7. twoway rcap lb ub foreign || scatter mean foreign, ylabel(, angle(h))
> xscale(r(-0.2 1.2)) xlabel(0 1, tlcolor(none) valuelabel) legend(off)
> subtitle("`which'") name(g`k', replace)
  8. }
  (output omitted)
. graph combine g1 g2 g3 g4
```

The big idea is to loop over variables, producing a graph for each variable to be put together by the final `graph combine`. But I chose to loop directly over integers 1 to 4, which I then used in naming the graphs. That choice is a small matter of style but also informed by experience. If I use similar code for similar problems, there is less code to revise if I cast the problem as a loop over integers rather than as a loop over variable names.

Know that `tokenize` parses its argument into words, parsing in this case a list of variable names on spaces. Then, the distinct words are assigned to local macros named 1 up. Thus, local macro with name `1` has contents `price`, and local macros with names `2`, `3`, and `4` have contents `mpg`, `weight`, and `length`, respectively. So that is how we can, at the same time, loop over 1 to 4 and the first, second, third, and fourth variable names, which are held within local macros named 1 to 4.

The trickiest detail here is the use of nested macro references, specifically ``` ``k'' ```. An easy analogy is with nested expressions in elementary algebra, such as $\{a-(b-c)\}$, where one quickly learns to evaluate first what is inside the innermost parentheses and then work outward. As local macro `k` loops over the values `1` to `4`, the nested reference ``` ``k'' ``` becomes in turn `` `1' ``, `` `2' ``, `` `3' ``, and `` `4' ``, and so a reference to the local macros `1` to `4`, which were earlier assigned as contents the names `price` through `length`.

Some of the details here are specific to the example. It seems to me good practice to show the units of price, here U.S. dollars, especially for an international readership. There are two distinct values of `foreign`, 0 and 1, but we want to see the corresponding value labels instead. Adding a little space on either side with `xscale()` is an aesthetic choice, as is suppressing the (visibility of the) tick by setting its color to none (Cox and Wiggins 2019).

The other details are more general. Each graph shows what it is about in a subtitle. For that subtitle, my preference is to use a variable label. All the variables here do in fact have variable labels, but the code also shows technique for using the variable name, rather than the variable label, when the latter is not defined.

Why did I use the `replace` in naming those graphs? We haven't used those names before in this tip. It is a gesture to human frailty. I almost never get the code exactly right first time round. There is always some detail I have forgotten, or got slightly or very wrong. Indeed, it is better to let Stata tell you quickly what needs changing. I started computing in the middle 1970s, when a job was submitted to the university computer (there was only one such) on a card deck with 80 column cards for data and code you punched yourself. A job would typically be run some hours after submission, and so there was much point to checking your cards repeatedly to try to get them exactly right, because one mistake would mean a delay of several hours before the next version could be tried. Now agonizing about code has less value when Stata, or some other program, will usually find the mistakes much faster than you can.

It is likely that you would prefer a graph scheme different from `sj`, perhaps one that you devised yourself. My personal favorite is `s1color`, although I often tweak away from its defaults.

It would have been entirely possible to write that code directly using variable names. Let's give the code as rewritten that way.

```
. set scheme sj
. foreach v in price mpg weight length {
  2. sysuse auto, clear
  3. if "`v'" == "price" label var price "Price (USD)"
  4. local which : variable label `v'
  5. if "`which'" == "" local which "`v'"
  6. statsby, by(foreign) clear: ci mean `v'
  7. twoway rcap lb ub foreign || scatter mean foreign, ylabel(, angle(h))
> xscale(r(-0.2 1.2)) xlabel(0 1, tlcolor(none) valuelabel) legend(off)
> subtitle("`which'") name(g`v', replace)
  8. }
  (output omitted)
. graph combine gprice gmpg gweight glength
```

Some of the choices here are just conventional, whether widespread (many people do it) or personal (it could just be me). Thus, names like `i`, `j`, `k` for looping macros that often start at 1 have been long used in many programming languages and echo even longer-standing mathematical practices. I often use the local macro name `v` in looping over variable names. If you wanted to use longer local macro names, only taste, how much you have to type, and Stata's limit of 31 characters for such names might stand in your way.

Some of that may have been new to you, or else you skimmed and skipped familiar constructs. To focus now on the main theme of this tip, let us see the result of `graph combine` in figure 1.

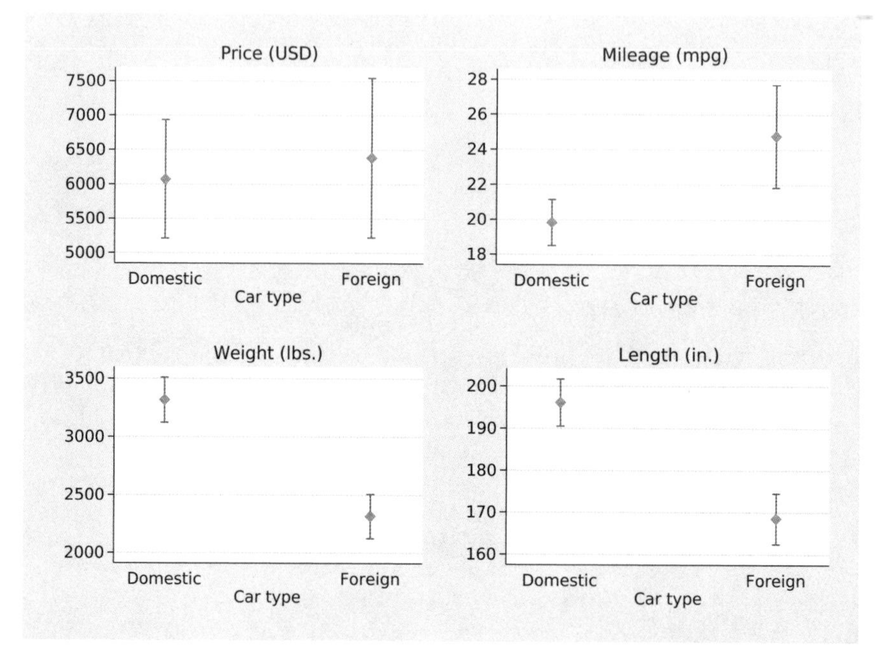

Figure 1. Means and confidence intervals for price, miles per gallon, weight, and length for domestic and foreign cars from the auto dataset. So far, so good, but the y axes stand out as using different spacing.

That does look quite good to me, but it could still be better. The different variables come with different units of measurement and—more crucially—different magnitudes, and the y axes are not exactly aligned. That is the niggling detail that people often spot quickly. Note also that specifying the `ycommon` option with `graph combine` solves this problem, but not helpfully, while specifying the `xcommon` option does no harm, but is no help either.

To solve this problem of unaligned axes, we turn instead to combining the reduced datasets and then drawing a graph just once using the `by()` option. This is more work, but the scope for improved appearance should make it seem worthwhile.

```
. tokenize "price mpg weight length"
. forvalues k = 1/4 {
  2. sysuse auto, clear
  3. if "``k''" == "price" label var price "Price (USD)"
  4. local which : variable label ``k''
  5. if "`which'" == "" local which "``k''"
  6. local labels `labels' `k' "`which'"
  7. statsby, by(foreign) clear: ci mean ``k''
  8. generate which = `k'
  9. save g`k', replace
 10. }
  (output omitted)
. append using g1 g2 g3
  (output omitted)
. label define which `labels'
. label values which which
. list, sepby(which)
  (output omitted)
. set scheme sj
. twoway rcap lb ub foreign || scatter mean foreign,
> by(which, yrescale note("") legend(off)) ylabel(, angle(h)) xscale(r(-0.2 1.2))
> xlabel(0 1, tlcolor(none) valuelabel)
```

The results of `list` are suppressed here, but having a look at the data is a really good idea if you are working through the code yourself (ideally to adapt the idea to some different dataset you care about).

We need to store information on each variable for later use as we loop through them. We want the variable labels of the four variables concerned (or if they do not exist, the variable names) to become value labels for the grouping variable we are going to use. Naturally, for that to be possible, the grouping variable, here called `which`, must be created first.

For this version of the problem, assigning integers 1 to 4 in the order in which we want the panels is definitely a good idea. Otherwise, if the variable names `price mpg weight length` are used as distinct values of `which`, and that variable is fed to `by()`, then those panels will be shown in alphabetical name order. The `by()` option will sort on the distinct values of the variables it sees, which here means alphabetical order for text that was originally variable names. (There is extra detail, which doesn't apply in this example, for underscore or numeric characters.) Alphabetical order has its uses (Flanders 2020), but alphabetical order of panels is rarely what you want, or more precisely, rarely what you should want. Wainer (1984, 1997) mocked unthinking use of alphabetical order in statistical graphics as Austria first! or Alabama first! Scottish readers will want to add Aberdeen first!

Figure 2 is the result. Isn't it better? What it shows is perhaps banal, but here we go: foreign cars typically are more expensive but have better mileage than domestic cars, and they are lighter and shorter on average too.

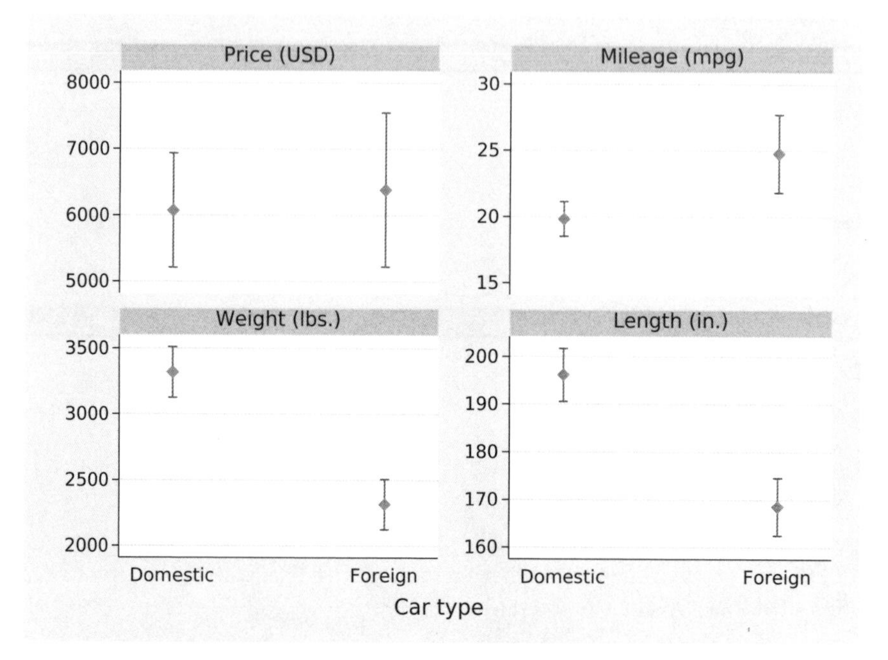

Figure 2. Means and confidence intervals for price, miles per gallon, weight, and length for domestic and foreign cars. It is not obvious from the graph, but a `by()` option was used to improve the presentation of y axes.

3 Example 2: Slicing cyclic time series

Effective graphical representation of intensely cyclical time series can be a challenge, especially if they are long. Many time series from several fields, including economics, epidemiology, and environmental science, are strongly seasonal. Cycles in those and other fields can also be identified with lengths both shorter and longer than a year. Cycles vary from fairly predictable to highly unpredictable, but the same graphical challenge is common to many: do justice to the structure of cycles, which may itself be changing over time, and also reveal any trends or other long-term changes. Stata-based discussion in Cox (2006, 2009) covers several of the graphical ideas that have been floated.

Sunspot numbers are a popular sandbox for enthusiasts in statistical graphics (for example, Cleveland [1993, 1994]; Wilkinson [2005]). Here I use annual averages downloaded on 3 June 2020 from http://www.sidc.be/silso/home with grateful acknowledgment to WDC-SILSO, Royal Observatory of Belgium, Brussels. The data are available with the files for this tip.

A standard line plot is not crazy but not especially helpful either (figure 3).

```
. use sunspots, clear
. line sunspots year
```

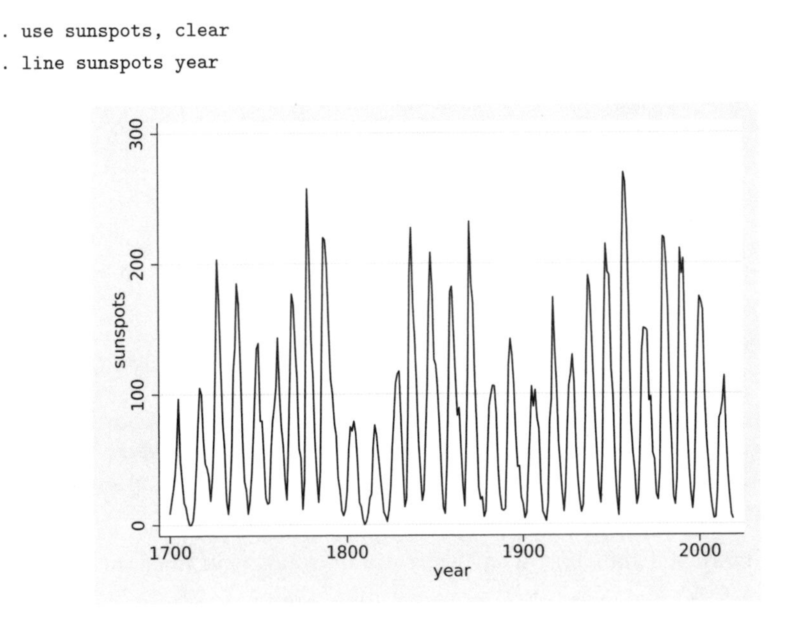

Figure 3. A common or garden line chart for mean sunspot numbers, 1700–2019. Strongly cyclical behavior is evident, but the fine structure of cycles is not especially clear from the roller-coaster display.

Common advice is to change the aspect ratio (ratio of graph height to graph width), which often can work well enough (Cox 2004). Changing the aspect ratio is sometimes discussed as if novel, but it was evidently familiar to Fisher (1925, 31) and doubtless earlier yet. Here the effect is a little disappointing. You may have found low aspect ratios used in long, concertina like graphs inserted in books, but changes in printing have made those much less common.

```
. line sunspots year, aspect(0.2)
```

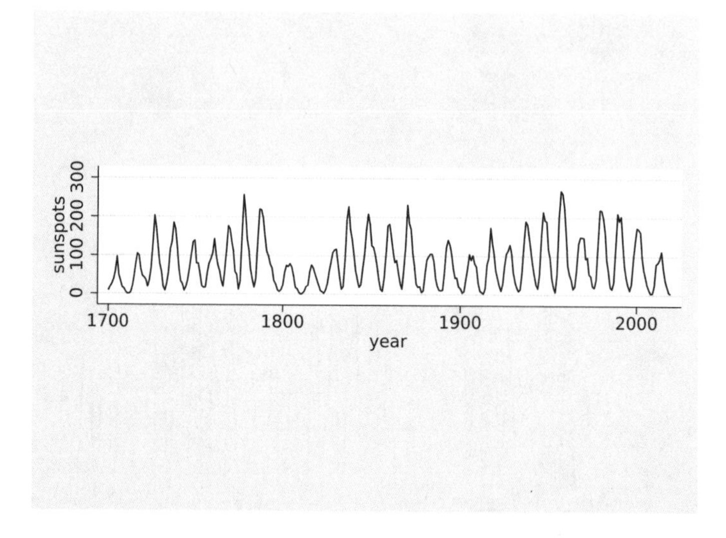

Figure 4. Changing the aspect ratio does not help much here

A twist on the same idea is to slice the series and use several panels for the different slices. Cleveland (1993, 1994) called these "cut-and-stack plots". Cox (2006) reported a Stata command `sliceplot`, but the novelty here is a belated recognition that often no such command is needed. Code comes first and then explanation.

```
. generate slice = ceil(4 *_n/_N)
. line sunspots year, by(slice, note("") cols(1) xrescale)
> ylabel(, angle(h)) xtitle("")
> subtitle("", position(9) nobox nobexpand fcolor(none))
```

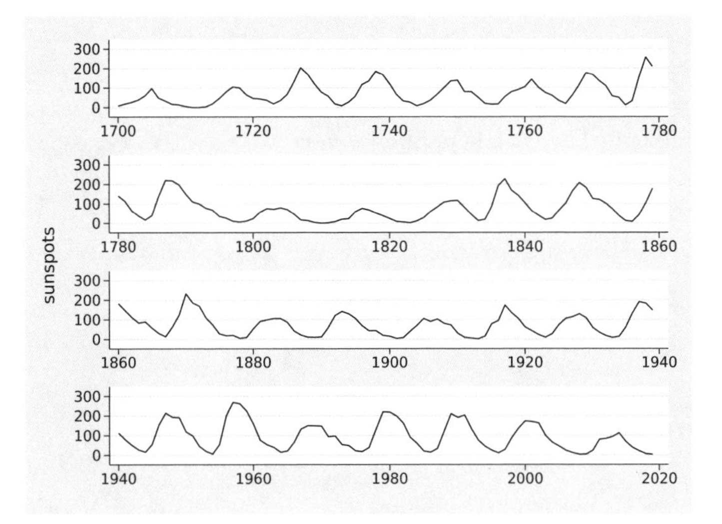

Figure 5. Slicing a long cyclical time-series plot makes the structure of the data clearer. Note the common asymmetry of cycles in which rising limbs are often steeper than falling limbs. A `by()` option was used to create this plot.

We create a slice variable that runs from 1 to 4, but naturally you should use whatever different value of 4 makes sense for your problem. (That recycles a mild joke associated with the great probabilist, William Feller, 1906–1970.)

To spell it out, the presumption is that the time series is already sorted on a time variable. The inbuilt observation number `_n` varies from 1 to the inbuilt number of observations in the dataset `_N`, so `4 * _n/_N` varies from almost 0 to 4. Rounding up with the `ceil()` function produces a variable containing a slice counter with integer values from 1 to 4.

That variable is just a means to an end, and so we suppress all evidence that it exists at all, blanking out both the `note()` that names it and the `subtitle()` that shows its distinct values. The `xrescale` suboption is essential for a readable graph: if you doubt that, see what happens if you omit it. Making the y-axis labels horizontal and removing the default `xtitle()` (which would be `year` in this example) are matters of taste. Aligning in one column of panels is a good idea to keep the aspect ratio low.

The payoff is not just in terms of a clearer graph. It becomes evident that cycles are often asymmetric, with steeper rising limbs than falling limbs. Explaining why is a problem for researchers in solar physics.

4 The art that conceals art

A duty to be clear is a higher virtue than using clever tricks, but clever tricks are always worth knowing about. The trick here implies more work by the user, and the payoff has to be judged case by case. It can be explained as art concealing art, in using a `by()` option but suppressing the visible evidence of doing it that way.

Further uses of the same device can be found in the community-contributed commands `multiline` (Cox 2017b) and `multidot` (Cox 2017a), downloadable from Statistical Software Components Archive (look at `help ssc` if that resource is new to you).

References

Cleveland, W. S. 1993. *Visualizing Data.* Summit, NJ: Hobart.

———. 1994. *The Elements of Graphing Data.* Rev. ed. Summit, NJ: Hobart.

Cox, N. J. 2004. Stata tip 12: Tuning the plot region aspect ratio. *Stata Journal* 4: 357–358. https://doi.org/10.1177/1536867X0400400313.

———. 2006. Speaking Stata: Graphs for all seasons. *Stata Journal* 6: 397–419. https://doi.org/10.1177/1536867X0600600309.

———. 2009. Stata tip 76: Separating seasonal time series. *Stata Journal* 9: 321–326. https://doi.org/10.1177/1536867X0900900211.

———. 2010. Speaking Stata: The statsby strategy. *Stata Journal* 10: 143–151. https://doi.org/10.1177/1536867X1001000112.

———. 2017a. multidot: Stata module for multiple panel dot charts and similar. Statistical Software Components S458376, Department of Economics, Boston College. https://ideas.repec.org/c/boc/bocode/s458376.html.

———. 2017b. multiline: Stata module for multiple panel line plots. Statistical Software Components S458369, Department of Economics, Boston College. https://ideas.repec.org/c/boc/bocode/s458369.html.

———. 2020. Speaking Stata: Loops, again and again. *Stata Journal* 20: 999–1015. https://doi.org/10.1177/1536867X20976340.

Cox, N. J., and V. Wiggins. 2019. Stata tip 132: Tiny tricks and tips on ticks. *Stata Journal* 19: 741–747. https://doi.org/10.1177/1536867X19874264.

Fisher, R. A. 1925. *Statistical Methods for Research Workers.* Edinburgh: Oliver and Boyd.

Flanders, J. 2020. *A Place for Everything: The Curious History of Alphabetical Order*. London: Picador.

Wainer, H. 1984. How to display data badly. *American Statistician* 38: 137–147. https://doi.org/10.1080/00031305.1984.10483186.

———. 1997. *Visual Revelations: Graphical Tales of Fate and Deception from Napoleon Bonaparte to Ross Perot*. New York: Copernicus.

Wilkinson, L. 2005. *The Grammar of Graphics*. 2nd ed. New York: Springer. https://doi.org/10.1007/0-387-28695-0.

 DOI: 10.1177/1536867X211000032

The Stata Journal (2021)
21, Number 1, pp. 263–271

Stata tip 140: Shorter or fewer category labels with graph bar

Nicholas J. Cox
Department of Geography
Durham University
Durham, UK
n.j.cox@durham.ac.uk

1 The problem: Messy category labels with graph bar

Stata does not have the concept of a categorical variable, but statistical people do, as shown by many book titles alone (for example, Fienberg [1980]; Lloyd [1999]; Simonoff [2003]; Tutz [2012]; Agresti [2013]; Long and Freese [2014]). The categories of such variables are distinct and often, but not necessarily, few in number. Examples are differing disease condition, employment status, or land cover type. A categorical variable in Stata can be stored as string; as numeric with value labels; or sometimes as just numeric, as when the number of cars or cats or children in households is just a discrete count that you want to treat as categorical.

The main focus of `graph bar` and its siblings `graph hbar` and `graph dot` is showing data or results for one or more outcomes and for distinct values of categorical predictor variables. Typically, the categorical predictors are named in the `over()` or `by()` option, or both. That way of thinking does not rule out using any of these commands in other ways. In practice, the difficulties discussed in this tip arise most commonly with `graph bar`, but if the solutions help with the other commands, that is fine. Also, those difficulties arise often with predictor variables that a researcher would not think of as categorical at all. The leading example here has that flavor.

The general problem addressed in this tip is that you are using `graph bar` and your categorical axis labels are a mess. They overlap and you need shorter labels, or fewer labels, or perhaps both. Although it may seem puzzling or even perverse, `graph bar` does not have an x axis; the horizontal axis is thought of as a categorical axis. The y axis is as usual the vertical axis showing outcome values or means or counts or whatever else the graph shows. One good reason for that idiosyncrasy is that typing `graph hbar` or `graph dot` instead flips the axes and indeed often improves the graph. The y axis remains the axis that shows the outcome, even though it is now horizontal. You would not want to have to edit a series of options naming x-axis properties to the corresponding y-axis properties, and conversely, which is what `graph twoway` often requires if you change axes.

The example in this thread of a bar chart for time series raises just about all the generic issues that arise commonly. It also raises some specific issues for time series that are frequent.

The immediate stimulus for this tip was a thread on Statalist,

> https://www.statalist.org/forums/forum/general-stata-discussion/general/1565227-graph-bar-over-year-how-to-shorten-displayed-year-labels

which started on 24 July 2020. R. Allan Reese in particular made one suggestion incorporated here.

Let's dive in and consider the use of a bar chart for two time series from the Grunfeld dataset bundled with your version of Stata. Bars side by side for two or more variables are often wanted and one of the attractions of **graph bar**.

We are setting on one side any discussion of whether some other kind of graph, such as a line chart, would be better; or of whether it is a good idea to juxtapose two series that may not have even the same units of measurement.

Time in the Grunfeld dataset is each year from 1935 to 1954, but **graph bar** does not know or care that one variable is time. The default sort order for time as a numeric categorical variable is almost always what you want anyway, but **graph bar** pays no attention to any gaps or unequal spacing of time values.

To set yet other details out of the way, let's say that we have decided in advance what bar colors we want and where the legend should go. In practice, a script for your own data may go through several iterations as you play with possibilities and discover small points to be resolved.

```
. webuse grunfeld
. set scheme sj
. local opts1 bar(1, fcolor(gs14) lcolor(black))
> bar(2, fcolor(gs3) lcolor(black))
. local opts2 legend(pos(11) ring(0) col(1))
. graph bar (asis) invest kstock if company == 1, over(year) `opts1´ `opts2´
```

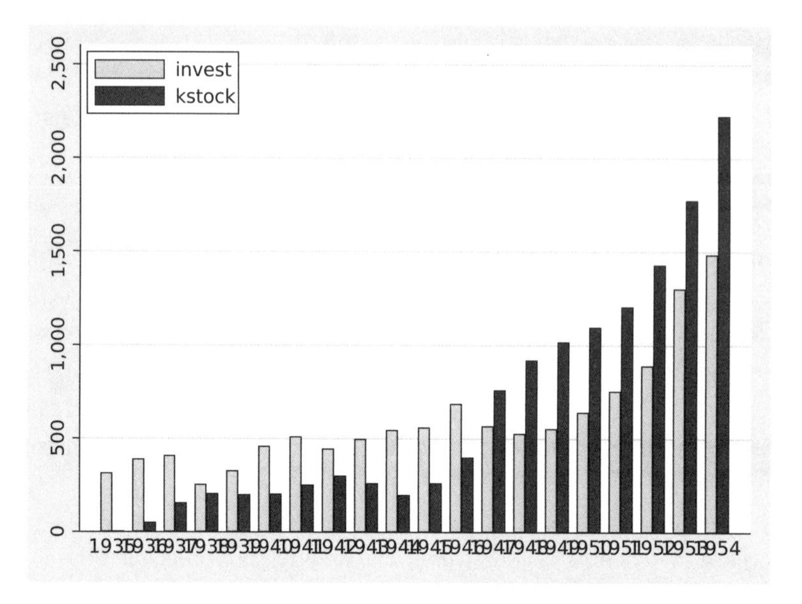

Figure 1. The time labels overlap. What to do?

Figure 1 is evidently poor, even though there are only 20 distinct years here. As you know, very many time series include far more times, but 20 is enough to cause a mess.

The thinking of `graph bar` is that you care about all your categories, so you want to know what each one is. But the default that every predictor value—every distinct category—is matched by an explicit text label is not always what you want for time series.

2 Possible solutions

2.1 Possible solution: Use graph hbar instead

Very often, the answer is simply to go horizontal by using `graph hbar`, not `graph bar`. Many of the vertical bar charts I see (some people call such charts "column charts") would be better off horizontal, giving space for longer, readable text labels and avoiding solutions that are awkward if not horrible, such as vertical labels, labels on a slant, overabbreviated labels, or labels in a tiny font size.

When this is the answer, good, and you bail out here.

For time series, this is not usually an acceptable answer. There is a strong convention across many fields that time belongs on the horizontal axis. Put your time variable on a vertical axis, and someone reviewing your work is likely to query that directly.

2.2 Possible solution: Use shorter text labels through value labels

You could assign value labels, such as "35" to 1935. `graph bar` respects value label definitions. You are in charge and are not obliged to use the same recipe throughout, so, for example, first and last value labels could be full length, say, "1935" or "1954".

Typing out definitions for 20 value labels (to say nothing of many more, as would be needed for longer series) is less fun than writing a loop to write code for your later `label` command. If you are new to loops, or indeed to local macros, my recent column (Cox 2020) provides a tutorial introduction.

```
. local call
. forvalues j = 1/20 {
  2.         local show = `j´ + 34
  3.         local year = 19`show´
  4.         local call `call´ `year´ "`show´"
  5. }
. label define year `call´
. label value year year
. graph bar (asis) invest kstock if company == 1, over(year) `opts1´ `opts2´
```

The initial statement

```
. local call
```

deserves comment. Thereby, I blank out any contents of the local macro call that exist (equivalently, I delete the macro if it exists). That way, I will not get bitten because of whatever is left behind in the macro from any previous code. This device will be used repeatedly in what follows. Here the macro does not exist at the outset, but explicitly blanking it out is suggested as good style.

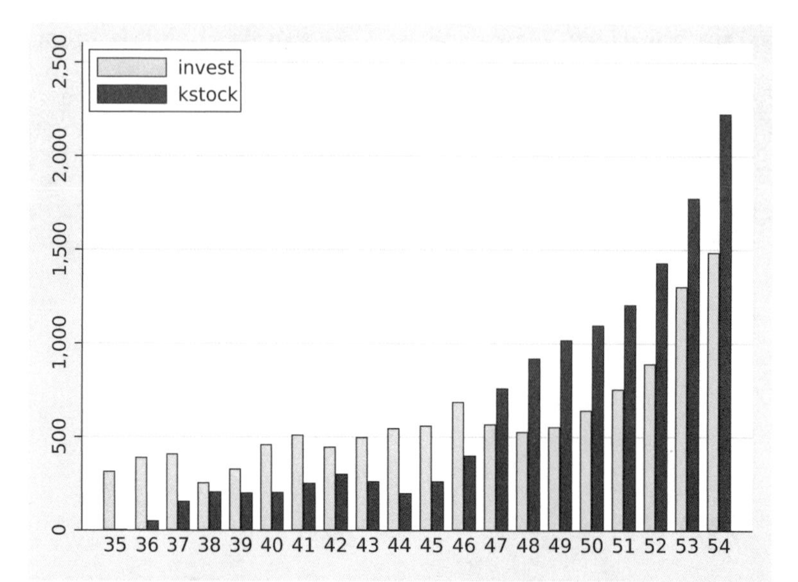

Figure 2. Insisting on shorter text labels (through value labels) removes the overlap. There are still questions: Do you want all those labels? Would this work with more distinct times?

Figure 2 is much better. You might want to stop here. But to spell out what is key: What we did was set up value labels such as `"35"` for 1935. Then, `graph bar` did not need to be told to look for and use value labels when they exist. Its expectation, as said, is that the variable named to `over()` is categorical. If that variable is numeric, then it is very likely to have value labels attached.

So far so good, but even for this example you might think "too many labels!". Your real data might be a time series with many more values, and if so, you really would think that.

Value labels that are just `"35"` to `"54"` for 1935 to 1954 are easy enough, but what if your years were 2000 to 2019 and you wanted the graph to show `"00"` to `"19"`? See a previous tip (Cox 2010) for how to get leading zeros when wanted.

Note that although you can set some of your value labels to spaces or even more exotic characters such as `char(160)` or `uchar(160)`, such value labels are not honored by `graph bar`. So we need some other device.

2.3 Possible solution: Use relabel()

The help for `graph bar` tells you about this suboption, enabling you to spell out whatever text you want. That does allow blanking out in a strong enough sense. In our running example, `graph bar` numbers the bars 1 to 20 (and so not 1935 to 1954). Here I write code using a loop for the suboption `relabel()` such that odd-numbered bars have as text a two-digit year (omitting century information) and even-numbered bars have as text a space (which you should not notice, except as resembling no text). Detail: an empty string will not work, because the command will not believe that you do not want a label at all.

```
. local call
. forvalues j = 1/20 {
  2.        local show = `j´ + 34
  3.        if mod(`j´, 2) local call `call´ `j´ "`show´"
  4.        else local call `call´ `j´ " "
  5. }
. graph bar (asis) invest kstock if company == 1, over(year, relabel(`call´))
> `opts1´ `opts2´
```

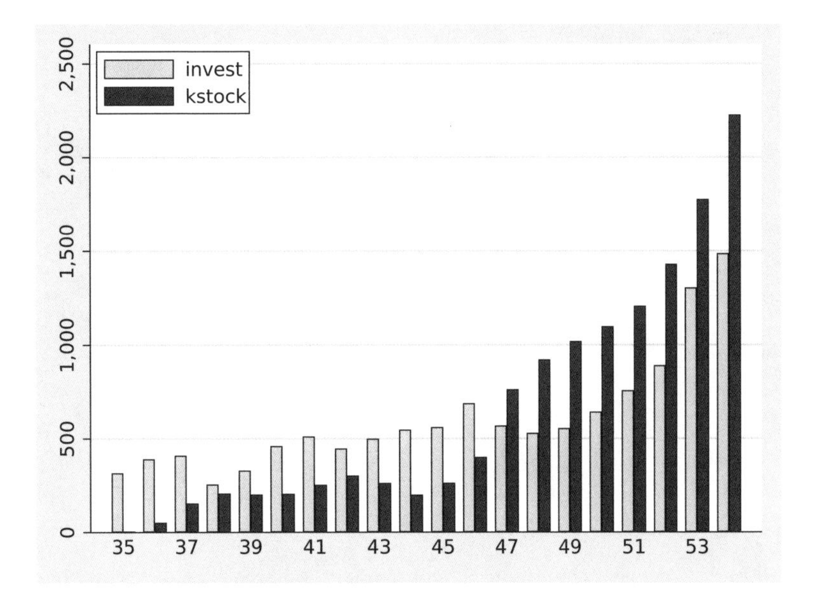

Figure 3. Every other year label is blanked out. Strictly, you are seeing spaces as value labels for even years, but the effect is identical.

Figure 3 shows the effect of showing fewer text labels.

If the `mod()` function is new to you, here is the small story. `mod()` returns the remainder on dividing its first argument by its second argument. (This is a long-standing abuse of terminology, because in mathematics the modulus is the divisor, not the remainder.) Division by 2 yields remainder 1 from odd integers and remainder 0 from even integers. A merit of this function is that it can be applied quite generally. A previous tip (Cox 2007) if anything understated that case. Indeed, functions in Stata are often undervalued by users (Cox 2011). So, in this example, if we wanted to show 1935 1940 1945 1950, these years have remainder 0 on dividing by 5. As a detail relevant to that example, note that `graph bar` will not let you show a label of 1955, because there is no bar for 1955 to label.

A guess suggests, and experiment confirms, that for these data we have enough space to show 1935 1937 to 1953. Let's clear the value labels out of the way and just show spaces instead of the even-numbered years.

```
. label val year
. local call
. forvalues j = 2(2)20 {
  2.        local call `call´ `j´ " "
. }
. graph bar (asis) invest kstock if company == 1, over(year, relabel(`call´))
> `opts1´ `opts2´
```

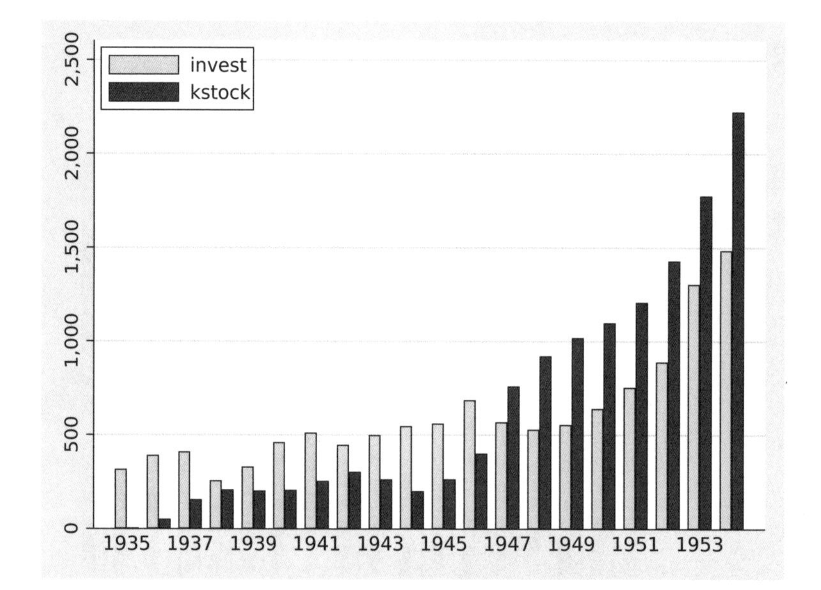

Figure 4. Showing every other year works well for these data. Strictly, the text for omitted years is a space, but the effect is as desired.

Figure 4 is in my view better than any previous figure. The same amount of text is shown on the x axis, but few readers should have difficulty imagining the omitted labels. There is not the puzzle of decoding 35 to 53 as years in their century.

2.4 Possible solution: Use twoway bar instead

You may be thinking that all of this is more messing around than you want for a very mundane problem. Is not there a simpler solution? For time series, there is, and it hinges on switching to `twoway bar`.

For bars side by side, you need to work a little at defining offset variables (Cui 2007). For two bars, we move one bar to the left and one bar to the right. We do need some small arithmetic to determine not just bar offsets but also good bar widths (which default to 1). Although we do not spell out any details, you could set up three or more bars, with the usual trade-off between the information gain in encoding several variables and the difficulty of decoding the several bars easily and effectively.

Now, your horizontal axis labels really are controllable simply and directly by the option `xlabel()`. Partly to show that we can, this final example uses axis labels every 5 years for simplicity, including the previously out-of-reach 1955:

```
. generate yearL = year - 0.15
. generate yearR = year + 0.15
. twoway bar invest yearL if company == 1, barw(0.3) fcolor(gs14) lcolor(black)
> || bar kstock yearR, barw(0.3) fcolor(gs3) lcolor(black) xla(1935(5)1955)
> `opts2´
```

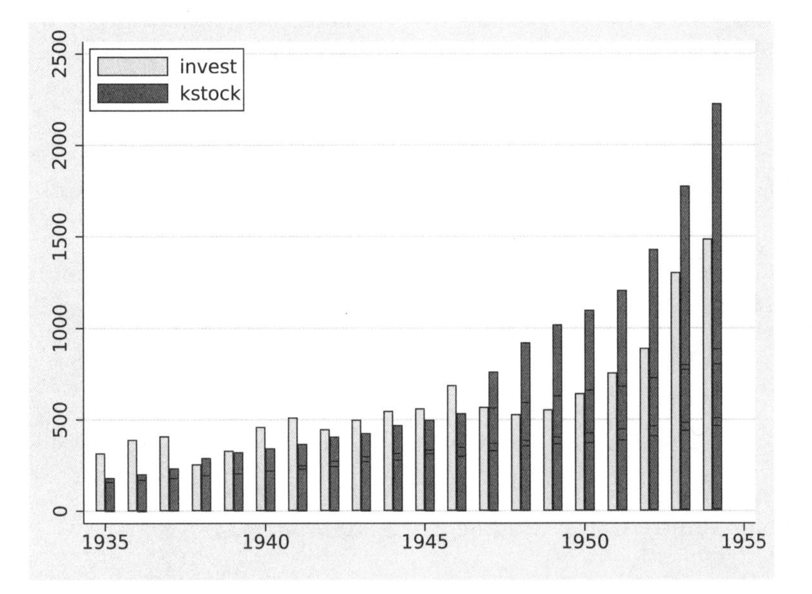

Figure 5. Just use `twoway bar` instead. You need to work at defining offset variables, but once that is done, it is downhill all the way.

Figure 5 plots one series of bars against one year variable offset left and the other series against one year variable offset right. The upshot is that `twoway` does not know what I want as x-axis title. That is fine by me, because I do not want to display an axis title like `"year"` at all. If I want or need to explain that series are plotted against year, I can do that in the text option I write in my text editor or word processor. Often enough, it will be clear from context. Your teachers were more right than wrong in urging you to explain every graph axis, but calendar year can be an honorable exception.

Many users start with `twoway` in any case for time series. If the time series has gaps or missing values, that is a really good idea. If you want line charts, unconnected or connected, it is essential.

2.5 Edit the value labels or string values

It is not relevant here, but often the solution is to slim down the existing value labels or string values. It can be prudent to save the original versions in case you want to go back for some later purpose.

References

Agresti, A. 2013. *Categorical Data Analysis*. 3rd ed. Hoboken, NJ: Wiley.

Cox, N. J. 2007. Stata tip 43: Remainders, selections, sequences, extractions: Uses of the modulus. *Stata Journal* 7: 143–145. https://doi.org/10.1177/1536867X0700700113.

———. 2010. Stata tip 85: Looping over nonintegers. *Stata Journal* 10: 160–163. https://doi.org/10.1177/1536867X1001000115.

———. 2011. Speaking Stata: Fun and fluency with functions. *Stata Journal* 11: 460–471. https://doi.org/10.1177/1536867X1101100308.

———. 2020. Speaking Stata: Loops, again and again. *Stata Journal* 20: 999–1015. https://doi.org/10.1177/1536867X20976340.

Cui, J. 2007. Stata tip 42: The overlay problem: Offset for clarity. *Stata Journal* 7: 141–142. https://doi.org/10.1177/1536867X0700700112.

Fienberg, S. E. 1980. *The Analysis of Cross-Classified Categorical Data*. 2nd ed. Cambridge, MA: MIT Press.

Lloyd, C. J. 1999. *Statistical Analysis of Categorical Data*. New York: Wiley.

Long, J. S., and J. Freese. 2014. *Regression Models for Categorical Dependent Variables Using Stata*. 3rd ed. College Station, TX: Stata Press.

Simonoff, J. S. 2003. *Analyzing Categorical Data*. New York: Springer. https://doi.org/10.1007/978-0-387-21727-7.

Tutz, G. 2012. *Regression for Categorical Data*. Cambridge: Cambridge University Press. https://doi.org/10.1017/CBO9780511842061.

The Stata Journal (2021)
21, Number 3, pp. 838–846 DOI: 10.1177/1536867X211045583

Stata tip 141: Adding marginal spike histograms to quantile and cumulative distribution plots

Nicholas J. Cox
Department of Geography
Durham University
Durham, UK
n.j.cox@durham.ac.uk

1 Marginal spike histograms

Some intriguing examples given by Harrell (2015) prompt experiment with adding marginal spike histograms to quantile and (empirical) cumulative distribution (function) plots. This tip explains why this might be helpful and gives sample code using the official command `quantile`.

The point is at least twofold.

- *Pedagogic.* Such additions help clarify what (for example) a quantile plot means and how it can be explained as stacking values in terms of their associated cumulative probabilities. Every kind of graph becomes comfortable only with increased familiarity. I have encountered students and colleagues evidently long past their first histogram who struggle a little on their first experience with quantile plots. A little help does no harm.
- *Practical.* In principle, marginal histograms add no more information to a quantile plot. In practice, they offer a complementary view of each distribution, affording another way to think about distribution level, spread, and shape—and indeed fine structure too.

Discussion here is phrased entirely in terms of official Stata commands, with the `quantile` command (see [R] **Diagnostic plots**) the main workhorse, but the idea extends easily to community-contributed (user-written) commands such as `qplot` or `distplot` from the *Stata Journal* (Cox [1999a, 1999b, 2004, 2005]; use `search` to find recent updates).

As usual, the idea has longer roots. See Galton (1889, 38) for (in modern terms) a quantile plot and a histogram sharing the same vertical magnitude axis.

2 Enhancing quantile plots

The name "quantile plot" goes back at least to Wilk and Gnanadesikan (1968), but the device is much older, being used several times in the 19th century and early 20th century. The main idea is just to plot ordered values, often but not invariably on the y axis, against their associated cumulative probabilities on the other axis. More generally, ordered values are plotted against corresponding quantiles of some relevant distribution. Small print aside, associated cumulative probabilities are just quantiles of a uniform (rectangular, flat) distribution on the interval from 0 to 1. Cumulative distribution plots are—again, often but not invariably—plotted with those axes reversed.

There are many small variations in both terminology and format. Quantile–quantile plot and empirical cumulative distribution function plot are some other terms. As if in apology or compensation for such long-windedness, concise but more cryptic labels such as q–q plots and ECDF plots can be found. As a matter of curious convention rather than compelling logic, quantile plots commonly use markers or point symbols for each distinct value, while cumulative distribution plots commonly use connected lines. That is a distinction without a difference if any distribution is such that points merge into lines, as may well be true with large sample sizes.

Let's start with a simple quantile plot. The official Stata command `quantile` adds a reference line that allows comparison with a uniform (rectangular, flat) distribution with the same range as the observed data. I almost never want that, so I usually make it invisible by changing its color. I also often vary the marker symbol and vertical axis label from the default. Figure 1 shows such a plot.

```
. sysuse auto
(1978 automobile data)
. set scheme sj
. quantile mpg, rlopts(lc(none)) ms(oh) yla(, ang(h))
```

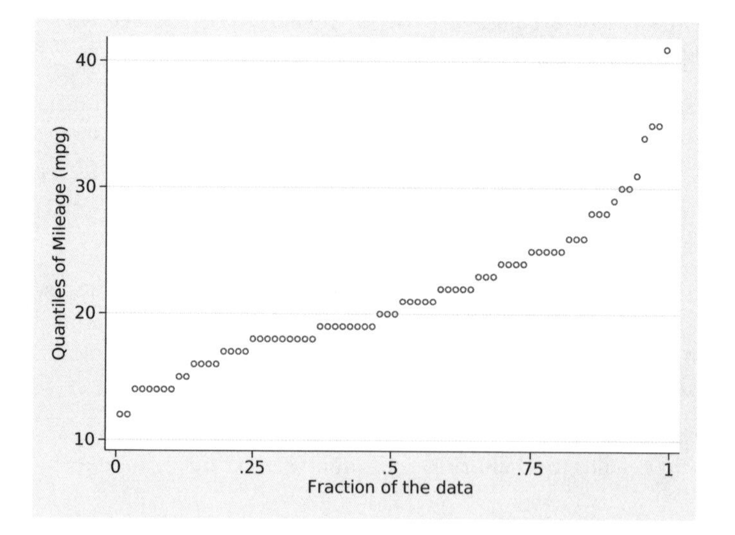

Figure 1. Quantile plot of miles per gallon from the auto data

As implemented in `quantile`, the idea is to plot values in rank order versus cumulative probability, specifically (unique rank − 0.5) / sample size. As explained in, say, Cox (2021), a rule rank / sample size does not treat lower and upper quantiles symmetrically and would cause other problems.

Because one axis is a probability scale, that is compatible in principle with a marginal histogram showing probabilities in each distinct bin. There are various fairly simple ways to get that directly. Here's one.

```
. count if mpg < .
  74
. egen prob = total(1/`r(N)'), by(mpg)
. egen tag = tag(mpg)
```

Counting the sample size explicitly does no harm and is needed if there are missing values (not the case in this example but often true) or if you want to look at a subset of your data (again, not the case here but also often true).

A first attempt just adds spikes to the vertical axis, as in figure 2.

```
. quantile mpg, rlopts(lc(none)) ms(oh) yla(, ang(h))
> addplot(spike prob mpg if tag, horizontal) legend(off)
```

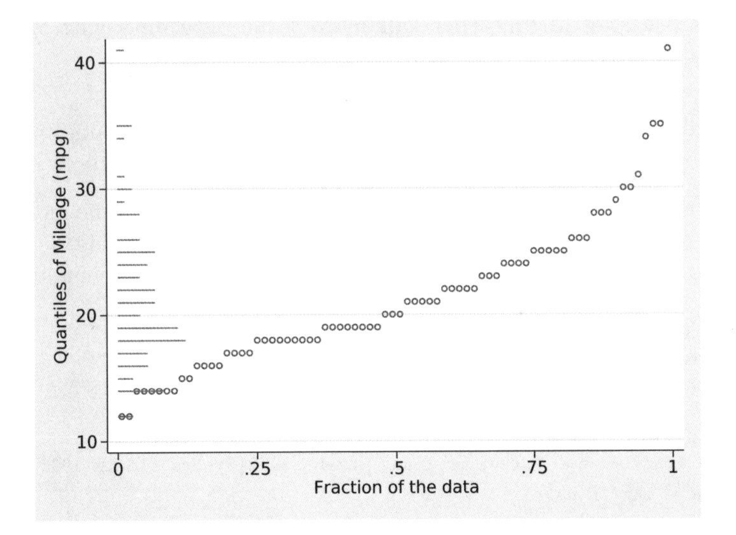

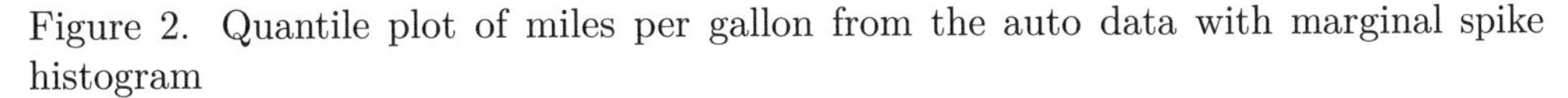
Figure 2. Quantile plot of miles per gallon from the auto data with marginal spike histogram

That is a good start, but we would be better off flipping the histogram so that it lies clear of the quantile plot. Negation is easy enough. See figure 3.

```
. generate nprob = -prob
. quantile mpg, rlopts(lc(none)) ms(oh) yla(, ang(h))
> addplot(spike nprob mpg if tag, horizontal) legend(off)
```

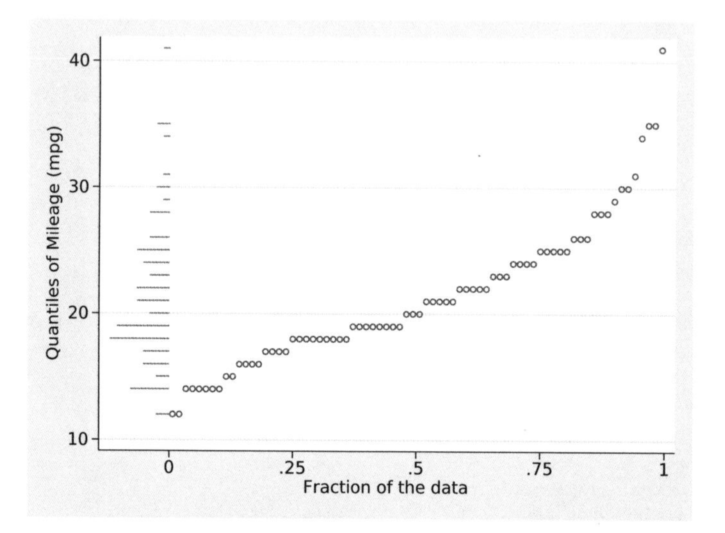

Figure 3. Quantile plot of miles per gallon from the auto data with marginal spike histogram, now flipped over or reflected in the vertical axis

The point about adding `if tag` is partly one of efficiency (the same spikes need not, and should not, be plotted repeatedly) and partly one of avoiding monitor artifacts.

In this example, we were lucky: There are 21 distinct values, so the mean probability in each bin is about 0.05. The maximum probability may naturally be much higher, but here taking probabilities literally (which means numerically) seems to work fine.

In other examples, we might have to do more work and make some decisions to get an extra histogram that is informative yet restrained. Let's look at a larger dataset. Figure 4 shows a distribution of wages on a natural logarithm scale.

```
. webuse nlswork, clear
(National Longitudinal Survey of Young Women, 14-24 years old in 1968)
. egen tag = tag(ln_wage)
. count if ln_wage < .
  28,534
. egen prob = total(1/`r(N)´), by(ln_wage)
. generate nprob = -prob
```

```
. quantile ln_wage, rlopts(lc(none)) ms(oh) legend(off)
> addplot(spike nprob ln_wage if tag, horizontal) yla(, ang(h))
```

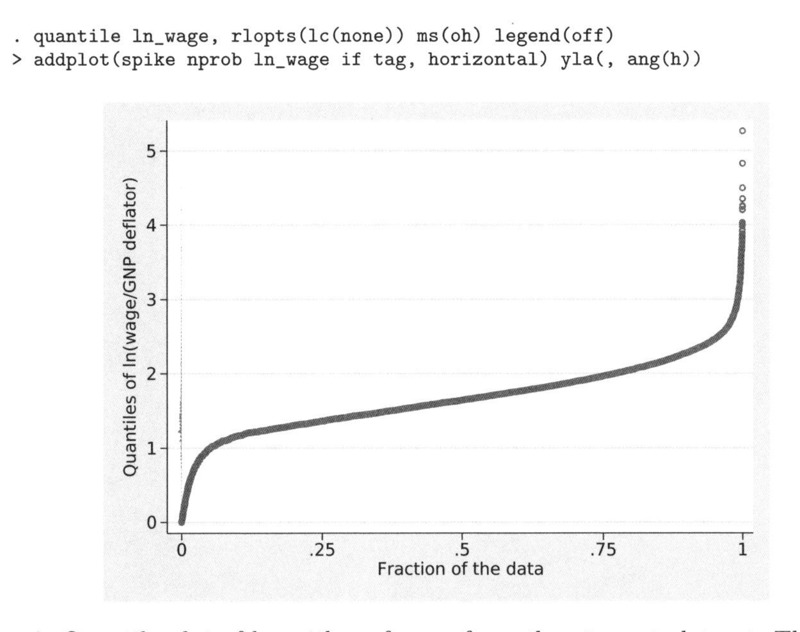

Figure 4. Quantile plot of logarithm of wage from the `nlswork` dataset. The marginal spike histogram is in effect a rug plot given the typically small bin probabilities.

Here there are 8,173 distinct values, and so the mean probability in each bin is about 0.0001. The maximum probability again can be much higher, but here taking probabilities literally (which means numerically) means that the histogram degenerates to a rug plot. If you want a rug plot, you can always get one (for example, Cox [2004]), but that is not the goal here.

There are various ways to move forward. One is to decide how much space to give the histogram and to scale accordingly. Then the probability scale for the quantile plot and that for the quantile plot are different, which needs to be explained somewhere. Figure 5 is an example. Another is to use a square-root scale for probabilities, which often works well to make structure clearer. Figure 6 is an example. The two ideas can be combined.

```
. summarize prob, meanonly
. generate nprob_scaled = nprob * (0.1 / r(max))
. generate prob_root = -sqrt(prob)
. quantile ln_wage, rlopts(lc(none)) ms(oh) legend(off)
> addplot(spike nprob_scaled ln_wage if tag, horizontal) yla(, ang(h))
. quantile ln_wage, rlopts(lc(none)) ms(oh) legend(off)
> addplot(spike prob_root ln_wage if tag, horizontal) yla(, ang(h))
```

Figure 5 scales the histogram to cover 10% of the horizontal extent of the quantile plot.

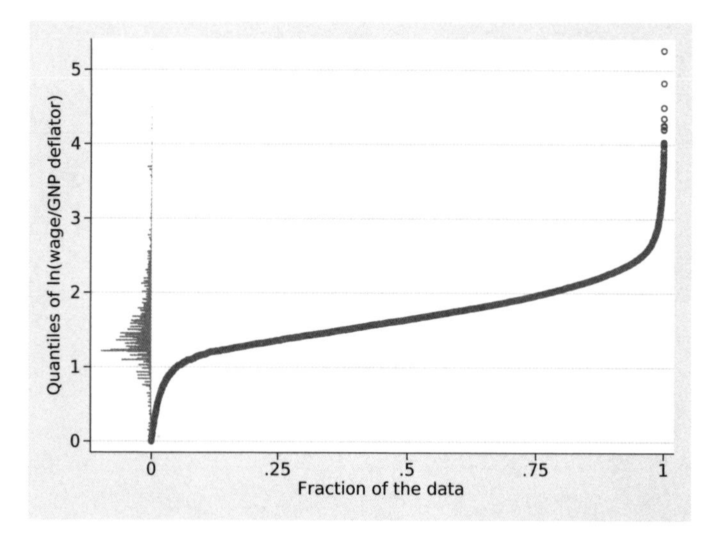

Figure 5. Quantile plot of logarithm of wage from the **nlswork** dataset. The spike histogram shows the same values but with a different probability scale.

Figure 6 shows a square-root scale.

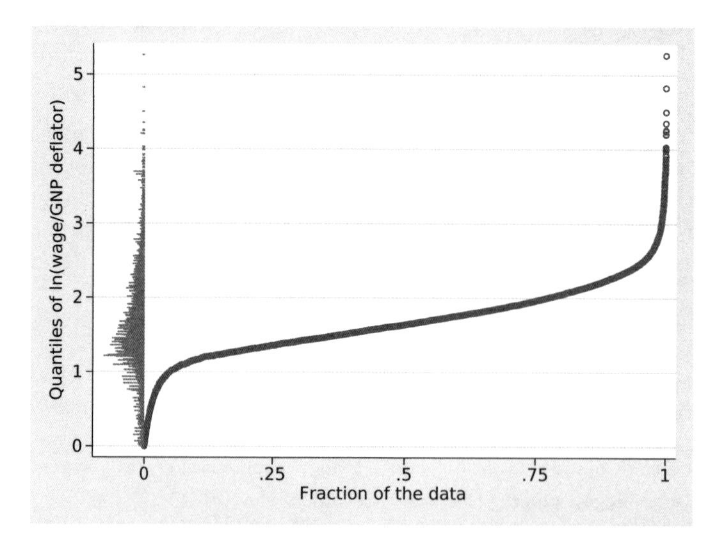

Figure 6. Quantile plot of logarithm of wage from the **nlswork** dataset. The spike histogram shows the same values but with a square-root probability scale.

3 Square-root scales for probability or frequency

If the idea of using a square-root scale is unfamiliar, let's spell that out. Square-root scales stretch back at least a century. See discussion and references in Cox (2012) and especially also Perrin (1913, 198; 1916, 131).

Histograms using square-root scales—rootograms!—go back at least to John W. Tukey circa 1965. The idea is to show not bin frequencies but their square roots. Frequencies, as counted variables, tend to have variability that is stabilized by a root transformation, at least approximately. It is now harder to see bins with higher frequency and much easier to see bins with lower frequency. Bins that are empty remain no problem because the square root of zero is zero. Note also that the square root of a normal or Gaussian density is a multiple of another normal or Gaussian density. Hence, if the normal is a reference distribution, we are looking for the same shape on a rootogram, and experience in assessing histograms for approximate normality can be applied directly in assessing rootograms. However, taking the root is only the first step in Tukey's procedure, and we do not implement his hanging or suspended rootograms. See Tukey (1986, 1972, 1977, chap. 17), Tukey and Wilk (1965), or Velleman and Hoaglin (1981).

4 A bimodal example

As suggested by Frank Harrell (personal communication), we close with an example that shows bimodal structure. We use the famous iris dataset of Edgar Anderson, as publicized by Fisher (1936) and often used as a sandbox in multivariate statistics and machine learning exercises. Stebbins (1978) gave an appreciation of Anderson, a distinguished and idiosyncratic botanist, and comments on the scientific background to distinguishing three species of the genus *Iris*. Kleinman (2002) surveys Anderson's graphical contributions with statistical flavor.

A Stata-readable copy of the iris data is bundled with the media for this issue. This dataset includes 150 measurements of the width and length of petals and sepals for 50 flowers each of the species *Iris setosa*, *Iris versicolor*, and *Iris virginica*. Using the same recipe as above, figure 7 shows the data for petal length.

The bimodal structure is, I suggest, clearly shown. It is no discovery and easily related to which species is being measured.

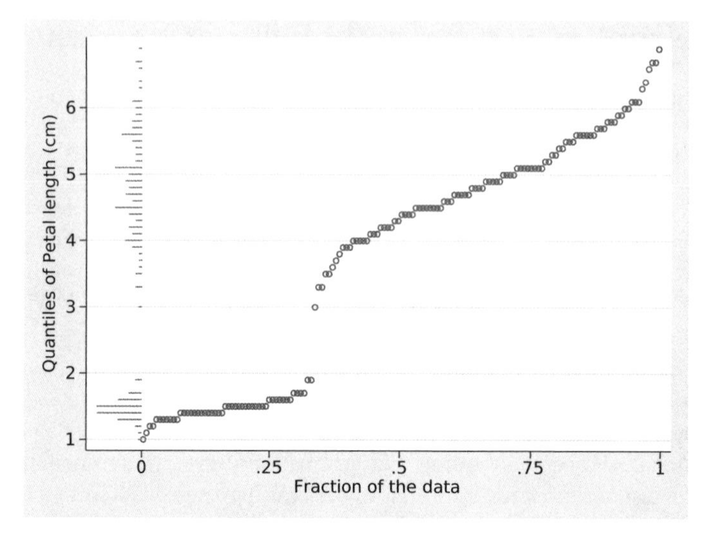

Figure 7. Quantile plot and marginal spike histogram for petal length from the iris data. Note the bimodal structure.

5 Graphics choices

In graphics, choices small and large abound, and taste and circumstance should determine your own choices. The device in this tip is for use whenever it is helpful. Other possibilities include offsetting the histogram slightly by introducing some space between it and the display of quantiles, using a marginal dot or strip plot instead, and smoothing the quantiles slightly because sometimes fine structure is just noise.

6 Acknowledgments

I thank Frank Harrell and Antony Unwin for encouraging and helpful correspondence.

References

Cox, N. J. 1999a. gr41: Distribution function plots. *Stata Technical Bulletin* 51: 12–16. Reprinted in *Stata Technical Bulletin Reprints*. Vol. 9, pp. 108–112. College Station, TX: Stata Press.

———. 1999b. gr42: Quantile plots, generalized. *Stata Technical Bulletin* 51: 16–18. Reprinted in *Stata Technical Bulletin Reprints*. Vol. 9, pp. 113–116. College Station, TX: Stata Press.

———. 2004. Speaking Stata: Graphing distributions. *Stata Journal* 4: 66–88. https://doi.org/10.1177/1536867X0100400106.

———. 2005. Speaking Stata: The protean quantile plot. *Stata Journal* 5: 442–460. https://doi.org/10.1177/1536867X0500500312.

———. 2012. Speaking Stata: Transforming the time axis. *Stata Journal* 12: 332–341. https://doi.org/10.1177/1536867X1201200210.

———. 2021. Speaking Stata: Front-and-back plots to ease spaghetti and paella problems. *Stata Journal* 21: 539–554. https://doi.org/10.1177/1536867X211025838.

Fisher, R. A. 1936. The use of multiple measurements in taxonomic problems. *Annals of Eugenics* 7: 179–188. https://doi.org/10.1111/j.1469-1809.1936.tb02137.x.

Galton, F. 1889. *Natural Inheritance.* London: Macmillan.

Harrell, F. E., Jr. 2015. *Regression Modeling Strategies: With Applications to Linear Models, Logistic and Ordinal Regression, and Survival Analysis.* 2nd ed. Cham, Switzerland: Springer. https://doi.org/10.1007/978-3-319-19425-7.

Kleinman, K. 2002. How graphical innovations assisted Edgar Anderson's discoveries in evolutionary biology. *Chance* 15(3): 17–21. https://doi.org/10.1080/09332480.2002.10554806.

Perrin, J. 1913. *Les Atomes.* Paris: Félix Alcan.

———. 1916. *Atoms.* New York: Van Nostrand.

Stebbins, G. L. 1978. *Edgar Anderson 1897–1969. Biographical Memoir.* Washington, DC: National Academy of Sciences. http://www.nasonline.org/publications/biographical-memoirs/memoir-pdfs/anderson-edgar.pdf.

Tukey, J. W. 1972. Some graphic and semigraphic displays. In *Statistical Papers in Honor of George W. Snedecor*, ed. T. A. Bancroft and S. A. Brown, 293–316. Ames, IA: Iowa State University Press.

———. 1977. *Exploratory Data Analysis.* Reading, MA: Addison–Wesley.

———. 1986. The future of processes of data analysis. In *The Collected Works of John W. Tukey.* Vol. *4, Philosophy and Principles of Data Analysis: 1965–1986*, ed. L. V. Jones, 517–547. Monterey, CA: Wadsworth and Brooks/Cole.

Tukey, J. W., and M. B. Wilk. 1965. Data analysis and statistics: Principles and practice. In *The Collected Works of John W. Tukey.* Vol. *5, Graphics: 1965–1985*, ed. W. S. Cleveland, 23–29. Pacific Grove, CA: Wadsworth and Brooks/Cole.

Velleman, P. F., and D. C. Hoaglin. 1981. *Applications, Basics, and Computing of Exploratory Data Analysis.* Boston: Duxbury.

Wilk, M. B., and R. Gnanadesikan. 1968. Probability plotting methods for the analysis of data. *Biometrika* 55: 1–17. https://doi.org/10.2307/2334448.

118 DOI: 10.1177/1536867X211063416

The Stata Journal (2021)
21, Number 4, pp. 1065–1068

Stata tip 142: joinby is the real merge m:m

Deni Mazrekaj
Department of Sociology
Utrecht University
Utrecht, The Netherlands
d.mazrekaj@uu.nl

Jesse Wursten
Faculty of Economics and Business
KU Leuven
Leuven, Belgium
jesse.wursten@kuleuven.be

1 merge versus joinby

The `merge` command is one of Stata's most used commands and works fine as long as the match key is unique in one of the datasets (that is, `merge 1:1`, `1:m`, or `m:1` situations). However, when the match key contains duplicates in either dataset, Stata gives an error message saying that `the key variable(s) do not uniquely identify observations in master or using dataset`. An example can clarify. In `jobs.dta`, we have two individuals. The first individual is a baker, and the second individual is a lawyer. The first and last names of each individual are included in `names.dta`.

names

	id	firstname	surname
1.	1	John	Smith
2.	2	Jane	Smith

jobs

	id	job
1.	1	baker
2.	2	lawyer

Merging these two datasets is straightforward in Stata:

```
. use "jobs", clear
. merge 1:1 id using "names", nogen
  (output omitted)
```

The `merge 1:1` shows that John Smith is a baker and Jane Smith is a lawyer.

```
. list
```

	id	job	firstn~e	surname
1.	1	baker	John	Smith
2.	2	lawyer	Jane	Smith

Things get tricky if we want to add information on the children of these workers. John and Jane Smith have two children together, Ken and Sue Smith.

children

	childfirstname	surname	age
1.	Ken	Smith	8
2.	Sue	Smith	6

`children.dta` can be linked to `names.dta` through their common surname, Smith. The Smith family consists of multiple parents and multiple children, suggesting we would need a `merge m:m`. However, this will not assign *both* children to *both* parents! Rather, `merge m:m` produces the nonsensical result below:

```
. merge m:m surname using "children", nogen
  (output omitted)
. list
```

	id	job	firstn~e	surname	childf~e	age
1.	1	baker	John	Smith	Ken	8
2.	2	lawyer	Jane	Smith	Sue	6

It linked the first parent to the first child and the second parent to the second child, rather than assigning both children to both parents.

Stata will not return an error message in this situation. Indeed, one motivation to write this tip was that we fear there are articles out there whose authors unwittingly ran a `merge m:m`. This is especially relevant in large population datasets that include millions of parents and children.

Instead, this situation requires the `joinby` command. Using the `joinby` command, we can add the information of each child to each parent.

```
. drop childfirstname age
. joinby surname using "children"
. list
```

	id	job	firstn~e	surname	childf~e	age
1.	1	baker	John	Smith	Sue	6
2.	1	baker	John	Smith	Ken	8
3.	2	lawyer	Jane	Smith	Ken	8
4.	2	lawyer	Jane	Smith	Sue	6

2 Stata example

Stata provides example `parent.dta` and `child.dta` datasets to test `joinby` yourself.

parent

family_id	parent_id	x1	x3
1025	11	20	643
1025	12	27	721
1026	14	26	668
1026	13	30	760
1030	15	32	684
1030	10	39	600

child

family_id	child_id	x1	x2
1025	3	11	320
1025	1	12	300
1025	4	10	275
1026	2	13	280
1027	5	15	210

We start by demonstrating the correct approach using `joinby`:

```
. webuse parent, clear
(Data on Parents)
. joinby family_id using "https://www.stata-press.com/data/r17/child"
. sort family_id parent_id child_id
```

The joined dataset correctly contains information on all parents (four) and on all their children (four in total). By default, `joinby` excludes unmatched observations. In this case, that means parents 10 and 15 and child 5 are not present in the joined dataset. This behavior can be changed through the `unmatched()` option.

```
. list, sepby(family_id)
```

	family~d	parent~d	x1	x3	child_id	x2
1.	1025	11	20	643	1	300
2.	1025	11	20	643	3	320
3.	1025	11	20	643	4	275
4.	1025	12	27	721	1	300
5.	1025	12	27	721	3	320
6.	1025	12	27	721	4	275
7.	1026	13	30	760	2	280
8.	1026	14	26	668	2	280

Just like `merge`, `joinby` also provides options to handle variables present in both datasets. By default, `joinby` retains the values of the primary dataset (compare `x1` above). Specifying the option `update` will update missing values with any nonmissing values in the secondary dataset. Adding the `replace` option will overwrite primary values with secondary values unless the latter are missing.

Contrast this result to the merge m:m outcome, where we specify keep(match) to mimic joinby's default of excluding unmatched observations.

```
. webuse parent, clear
(Data on Parents)

. merge m:m family_id using "https://www.stata-press.com/data/r17/child", nogen
> keep(match)

    Result                           Number of obs
    ---------------------------------------------
    Not matched                                 0
    Matched                                     5
    ---------------------------------------------

. sort family_id parent_id child_id
```

The merge m:m result below is extremely dangerous. In the first family (ID 1025), it assigned the first child (ID 3) to the first parent (ID 11) and all remaining children to the second parent (ID 12). In the second family, however, it correctly assigned the single child (ID 2) to both parents (IDs 13 and 14). As a result, if you manually checked the second family's results, you would assume merge m:m did exactly what you wanted even though it used a different allocation rule for the first family.

```
. list, sepby(family_id)
```

	family~d	parent~d	x1	x3	child_id	x2
1.	1025	11	20	643	3	320
2.	1025	12	27	721	1	300
3.	1025	12	27	721	4	275
4.	1026	13	30	760	2	280
5.	1026	14	26	668	2	280

The examples above highlight that you should (almost) never use merge m:m when the key match variable does not uniquely identify observations in both the master and the using datasets. Instead, if you want to assign all observations of the using dataset to all observations in the master dataset (for example, all children to all their parents), then joinby does exactly that.

The Stata Journal (2021)
122 DOI: 10.1177/1536867X211063414 **21**, Number 4, pp. 1069–1073

Stata tip 143: Creating donut charts in Stata

Andrew Musau
Molde University College
Molde, Norway
amus@himolde.no

Pie charts are circular charts cut by radii into segments illustrating relative magnitudes or frequencies. In Stata, one can create this graphic using the `graph pie` command (see [G-2] **graph pie**). A closely related graphic is the donut chart, essentially a pie chart with an area of the center cut out. The argument for pie charts or donut charts seems simple and can be summarized into the following three points:

i. They are widely used and widely familiar. In many places, pie charts are taught at very early ages. Familiar beats unusual, with nothing else said.

ii. The general idea of subdividing a total is very easy to grasp with a pie chart. Everyone can think of slices of a pie, or cake if you prefer, even small children or senior managers.

iii. The specific logic of a pie chart is that slice (more professionally, sector) angle, area, and arc length on the circumference are all proportional to quantity shown. The exception would be if pie slices are themselves divided radially, which gets very messy very quickly.

Conversely, the arguments against these graphics boil down to one argument: that other designs, bar or dot charts, work better and are not more difficult to understand. For a review of bar and dot charts and Stata techniques for producing them, see Cox (2008).[1] As of Stata 17, donut charts are not officially supported. However, there is some demand for this graphic by Stata users, as evidenced by posts on Statalist. Here we show that by first creating a pie chart, one can subsequently retrieve code generated from the Graph Editor (see [G-1] **Graph Editor**) to create a donut chart. Going through this process will not be necessary, because the code that I present here will directly create the chart, but the intermediate steps are set forth for replicability. For illustration, we will use `census.dta`, where we show proportions of population across census regions of the United States.

```
. sysuse census
(1980 Census data by state)
. graph pie pop, over(region) scheme(sj) legend(off)
```

1. I thank Nick Cox and an anonymous referee for detailed and constructive comments and suggestions.

The illustration is based on Windows 10 (Stata 17) but can be adapted for other operating systems. After running the code creating the pie chart, do the following:

1. Right-click on the pie chart, and click on **Start Graph Editor**.
2. On the Standard Toolbar, click on the **Start recording** button.

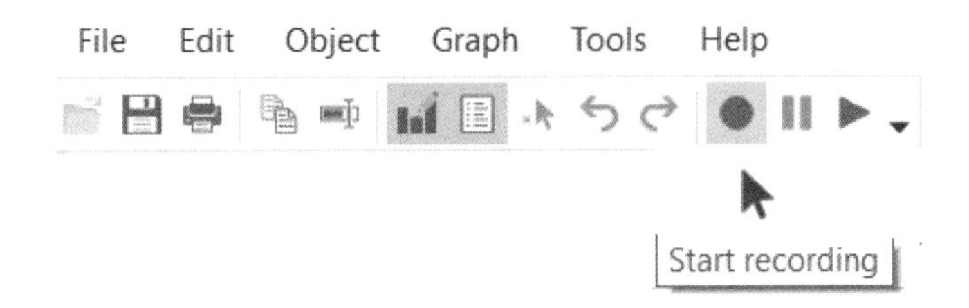

3. On the Tools Toolbar, click on **Add marker** button.

4. On the Contextual Toolbar, choose `White` as the *Color* and `Circle` as the *Symbol.*

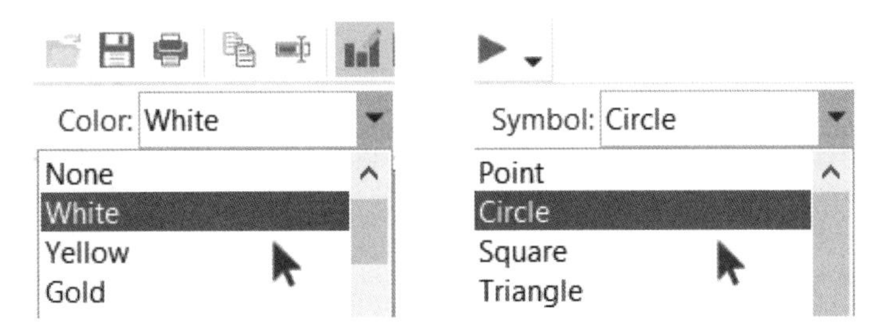

5. Place the marker at the center of the pie chart by placing the cursor there and clicking.
6. On the Standard Toolbar, click on the **End recording** button (same button as **Start recording** in number 2).
7. You will be prompted to provide a *Recording name:*, and you can choose where you want it to be saved.

The maximum area of the inner circle that can be specified using the Graph Editor is about a tenth of the area of the pie chart, not sufficient to create a donut chart. Therefore, we must modify the code retrieved from the Graph Editor to increase the inner circle's area. The recording is saved with the file extension `.grec`. On a Windows PC, use Notepad to open and access the code. The undocumented command `gr_edit` can be used to run lines of code retrieved from the Graph Editor. Below, we present a do-file code where we specify the size of the inner circle as a percentage of the pie chart's area in a local macro named `csize`. Note that a macro is simply a container holding text and that the preceding word `local` implies that the macro exists solely within the do-file or program in which it is defined (see [U] **18 Programming Stata** and Cox [2020] for an introduction to macros in Stata). Because of the existence of the local macro and the length of the code derived from the Graph Editor resulting in multiple line breaks, it is necessary to always run the entire do-file code in one go.

```
. local csize= 70
. gr_edit .plotregion1.AddMarker added_markers editor
> .0715713654788885 -.0361049561829105
. gr_edit .plotregion1.added_markers_new = 1
. gr_edit .plotregion1.added_markers_rec = 1
. gr_edit .plotregion1.added_markers[1].style.editstyle  marker( symbol(circle)
> linestyle( width( sztype(relative) val(.2) allow_pct(1)) color(white)
> pattern(solid) align(inside)) fillcolor(white) size( sztype(relative)
> val(`csize´*(93/100)) allow_pct(1)) angle(stdarrow) symangle(zero)
> backsymbol(none) backline( width( sztype(relative) val(.2) allow_pct(1))
> color(black) pattern(solid) align(inside)) backcolor(black)
> backsize( sztype(relative) val(0) allow_pct(1)) backangle(stdarrow)
> backsymangle(zero)) line( width( sztype(relative) val(.2) allow_pct(1))
> color(black) pattern(solid) align(inside)) area( linestyle( width(
> sztype(relative) val(.2) allow_pct(1)) color(ltbluishgray) pattern(solid)
> align(inside)) shadestyle( color(ltbluishgray) intensity(inten100)
> fill(pattern10))) label( textstyle( horizontal(center) vertical(middle)
> angle(default) size( sztype(relative) val(2.777) allow_pct(1)) color(black)
> position() margin( gleft( sztype(relative) val(0) allow_pct(1))
> gright( sztype(relative) val(0) allow_pct(1)) gtop( sztype(relative)
> val(0) allow_pct(1)) gbottom( sztype(relative) val(0) allow_pct(1)))
> linestyle( width( sztype(relative) val(.2) allow_pct(1)) color(black)
> pattern(solid) align(inside))) position(6) textgap( sztype(relative)
> val(.6944) allow_pct(1)) format(`""´) horizontal(default) vertical(default))
> dots( symbol(circle) linestyle( width( sztype(relative) val(.2) allow_pct(1))
> color(black) pattern(solid) align(inside)) fillcolor(black)
> size( sztype(relative) val(.1) allow_pct(1)) angle(horizontal) symangle(zero)
> backsymbol(none) backline( width( sztype(relative) val(.2) allow_pct(1))
> color(black) pattern(solid) align(inside)) backcolor(black)
> backsize( sztype(relative) val(1.52778) allow_pct(1)) backangle(horizontal)
> backsymangle(zero)) connect(direct) connect_missings(yes) editcopy
```

The graph obtained after running the do-file code from the previously created pie chart is shown in figure 1.

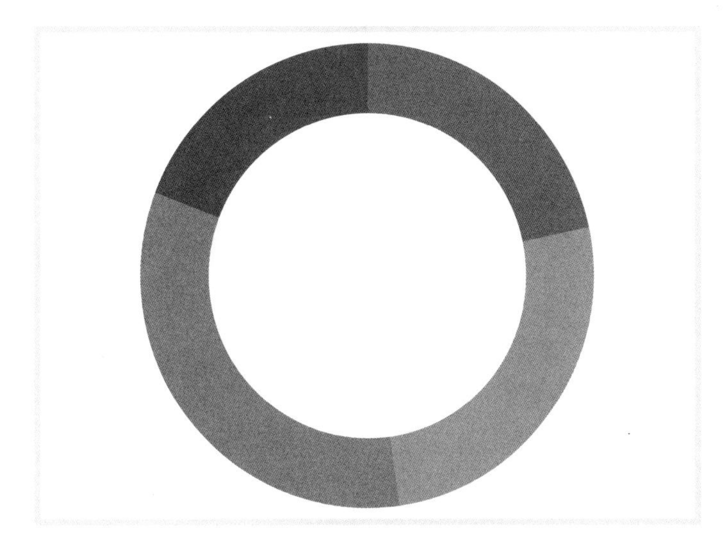

Figure 1. Donut plot without proportions and labels displayed

Because the inner circle is placed on top of the pie chart, it becomes apparent that proportions and labels must be placed outside the inner circle within the pie chart or completely outside the pie chart to be visible. Stata allows one to explicitly include either category labels or proportions. If proportions are included, category labels are shown in the legend, and if category labels are included, then proportions are not shown and one forgoes the legend.

Figure 2 illustrates a variant of each case. The `gap()` suboption specified within `graph pie`'s `plabel()` option determines the additional radial distance for the labels to appear on the pie chart's slices. When combining charts, note that the command `graph combine` (see [G-2] **graph combine**) does not shrink the sizes of the individual donut charts' inner circles; therefore, depending on how many charts need to be combined, one has to specify an appropriate value for the local `csize` in the do-file code. For the pie charts creating figure 2, we specify `csize=60`, just over 10% less than the value specified for figure 1.

```
. graph pie pop, over(region) scheme(sj)
> legend(ring(0) pos(2) bmargin(zero) col(1))
> plabel(_all percent, gap(10) color(white) size(medlarge))
. * run do-file code, specifying csize=60
  (output omitted)
. graph save gr1, replace
(file gr1.gph not found)
file gr1.gph saved
```

```
. graph pie pop, over(region) scheme(sj)
> legend(ring(0) pos(5) bmargin(zero) col(1))
> plabel(_all percent, gap(22) size(medlarge))
. * run do-file code, specifying csize=60
  (output omitted)
. graph save gr2, replace
(file gr2.gph not found)
file gr2.gph saved
. graph combine gr1.gph gr2.gph, scheme(sj)
```

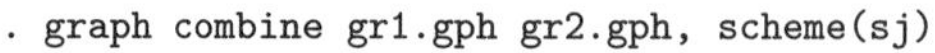

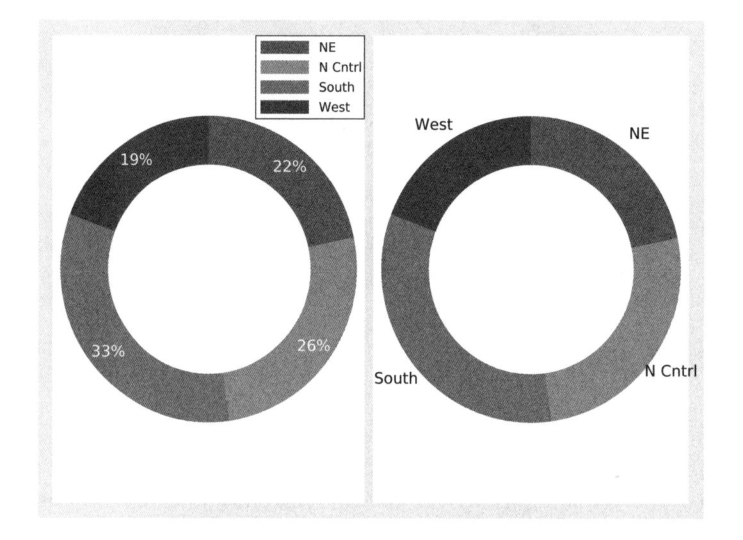

Figure 2. Combined donut plots with legends and proportions displayed

Having both proportions and category labels displayed requires that one insert each manually using the `ptext()` option.

References

Cox, N. J. 2008. Speaking Stata: Between tables and graphs. *Stata Journal* 8: 269–289. https://doi.org/10.1177/1536867X0800800208.

———. 2020. Stata tip 138: Local macros have local scope. *Stata Journal* 20: 499–503. https://doi.org/10.1177/1536867X20931028.

The Stata Journal (2021)
21, Number 4, pp. 1074–1080 DOI: 10.1177/1536867X211063413

Stata tip 144: Adding variable text to graphs that use a by() option

Nicholas J. Cox
Department of Geography
Durham University
Durham, UK
n.j.cox@durham.ac.uk

1 The problem

Graphs may be enhanced using a variety of title options, as explained in the help for `title options`. When the `graph` command also uses a `by()` option naming a variable to categorize data, the `subtitle()` automatically shows the value on that variable (or the associated value label if defined) of the particular group of observations being shown in each panel.

You might wish to add further text that varies from panel to panel. In particular, you might wish for scope to vary text in a `note()` or `caption()` from panel to panel. At the time of writing (Stata 17), that scope is not available. Either a note or caption applies to the entire display, or (less helpfully) the same note or caption is repeated for each panel.

This tip explains two alternative methods for adding text that varies informatively. A third method, not illustrated here, is to produce individual graphs as desired and then use `graph combine`. The advice here is a sequel or complement to an earlier tip (Cox 2011), which pointed out that title and legend options can be moved away from their default positions and placed within the plot region.

Our first example concerns how to add information on the number of observations to a scatterplot (figure 1).

```
. sysuse auto, clear
(1978 Automobile Data)
. set scheme sj
. scatter mpg weight, by(foreign, note(""))
```

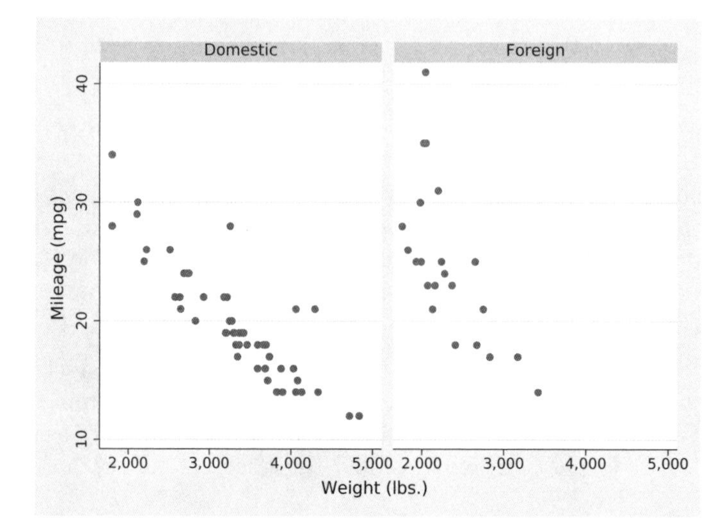

Figure 1. This is a simple example of a scatterplot using `by()`. How do we add extra text in each panel, say, flagging the number of observations in each group?

In all this, there is a permanent tension between making a graph more informative and wanting to avoid crowding and clutter. Much good advice (for example, Cleveland [1994]; Wilke [2019]) is on its face contradictory. It is a good idea to employ direct labeling, showing text right by data points or lines, rather at some distance as part of a key or legend. It is also a good idea to show the data clearly without congestion from extra text, so that the plot region is easy to scan and scrutinize. Circumstances and tastes affect choices, and the methods here are part of the flexibility you have as a graph designer.

2 Solution 1: Use an extra variable containing text as a marker label

We could create an extra string variable containing the text we need and then show that as a marker label. The marker label does not have to be associated with any variable already shown on the plot. On the contrary, it is often better to place the text well away from data points or lines in the plot region.

Looking at the graph in figure 1 suggests that a good location would be at `mpg` = 11, `weight` = 1800. (In practice, you may need some experimentation to find a position that seems good.) In our example, there are no missing values associated with `mpg` or `weight`, so we can get the number of observations shown from the number of observations `_N`, which under `by:` is determined groupwise. (See, for example, Cox [2002] on `by:` if all this is new to you.)

```
. generate y = 11
. generate x = 1800
. bysort foreign: generate toshow = "{it:n} = " + strofreal(_N)
```

Had there been any missing values, a more general technique would be to count the number of observations with nonmissing values:

```
. bysort foreign: egen count = total(!missing(mpg, weight))
. generate toshow = "it:n = " + strofreal(count)
```

The reasoning here: `!missing(mpg, weight)` returns 1 if both `mpg` and `weight` are not missing—the condition that defines an observation to be shown on the scatterplot—and 0 otherwise. Hence, adding results from that function across groups of observations puts the counts in a new variable.

An important flag: The Stata Markup and Control Language annotation instructing `graph` to show n in italic points up facilities for markup, including not just italic and bold fonts but also often-needed elements such as mathematical and other symbols, Greek characters, and subscripts and superscripts. See the help on `SMCL` and on `text` for the details.

We can now draw our graph. We are superimposing a plot of `y` versus `x`—which does just one thing, show extra text—on the original plot of `mpg` versus `weight`. A detail to fix is that with two y variables and two x variables, `graph` does not know what we want as axis titles. So here we reach in and insist on showing the variable labels for `mpg` and `weight`. Had no variable labels been defined, we could have used the variable names. In either case, we could have specified any other text that appeals for the axis titles. Similarly, we suppress the legend that would otherwise spring into existence: two y variables are being shown, but we do not need a legend entry for either. Figure 2 is the result.

```
. scatter mpg weight, ms(Oh) by(foreign, legend(off) note(""))
> || scatter y x, ms(none) mla(toshow) mlabsize(medium)
> ytitle("`: var label mpg´") xtitle("`: var label weight´")
```

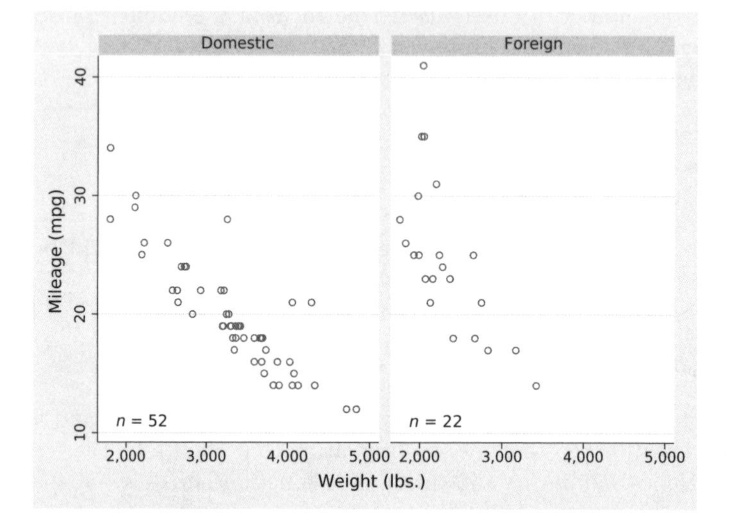

Figure 2. A string variable shown as marker labels is here used to add text in each panel indicating the number of observations shown. The value of the string variable is constant in each group, and variable from group to group, so the text shown matches that.

3 Solution 2: Modify the values or value labels of the by() variable

A second solution is to add information that will appear in the subtitles. Let's go through the possibilities:

1. If the `by()` variable is numeric without value labels, you will need to define value labels or a string variable to use instead.

2. If the `by()` variable is numeric with value labels attached, you will need to change the value labels or define a string variable to use instead.

3. If the `by()` variable is string, you will need to change the string variable or define a numeric variable with value labels.

The choice is a small deal, but consider carefully any implications for the order of the panels. As I flagged in a recent tip (Cox 2020b), alphabetical order of panels is rarely ideal. Wainer (1984, 1997) mocked unthinking use of alphabetical order in statistical graphics as Austria first! or Alabama first! Scottish readers will want to add Aberdeen first! Here we will use `decode`, which takes the value labels of `foreign` and makes them values of a string variable: the resulting order of `"Domestic"` and `"Foreign"` is, fortuitously but fortunately, exactly that of the original numerical values. At worst, the order would have been reversed, which would not have been a major problem— unless you were drawing other graphs with the same data and ended with displays using different conventions. But, in general, watch out: you will often need to define a new variable with a desired ordering that is numerically explicit.

Once we have `origin` as a string variable, we can add extra text. The new variable can be used in the `by()` option. Figure 3 is the result.

```
. decode foreign, gen(origin)
. bysort origin: replace origin = origin + " ({it:n} = " + string(_N) + ")"
(74 real changes made)
. scatter mpg weight, ms(Oh) by(origin, note(""))
```

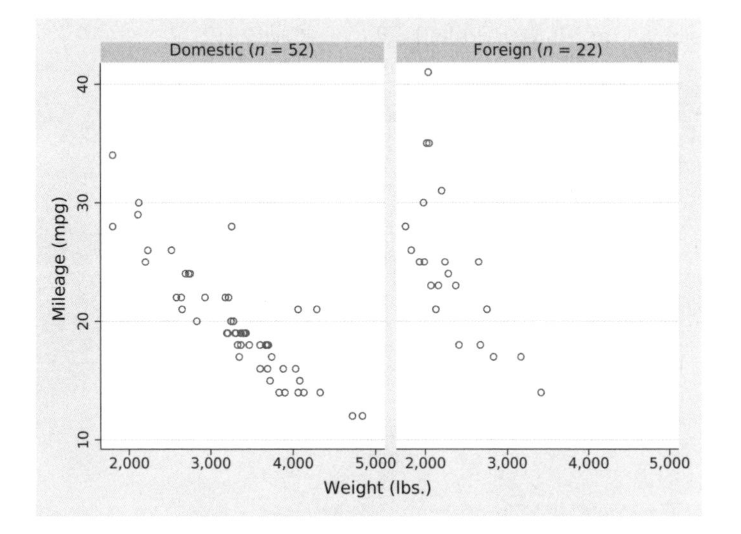

Figure 3. The text for the `by()` variable has been enhanced for this display. The result appears automatically as subtitles.

That was suspiciously easy, so let's close with a twistier example. Imagine evocative value labels for the prosaically numeric repair record variable `rep78`.

```
. label define rep78 1 abysmal 2 awful 3 adequate 4 admirable 5 outstanding
. label values rep78 rep78
```

Alphabetically ordered, the labels would not match the underlying numeric order, even though they come awfully close. So `decode` would mangle the order, and the nettle to be grasped is to redefine the value labels. Here is how to do this through syntax. Loop over the possible values, look up the value label, count the number of values, and then apply an augmented label. The syntax for looking up value labels is documented at `help macro`. The result of a `count` is stored temporarily in `r(N)`. Loops and local macros were introduced in a previous column (Cox 2020a).

```
. forvalues j = 1/5 {
  2.         local label : label (rep78) `j´
  3.         count if rep78 == `j´
  4.         local label "`label´ ({it:n} = `r(N)´)"
  5.         label define rep78 `j´ "`label´", modify
  6. }
  2
  8
  30
  18
  11
```

Then, we can use the new labels in a graph (figure 4). Here we go beyond the main theme of a `by:` option to show that the technique of enhancing value labels has wider applications. Personal taste shows in putting the information for the numeric axis at the top whenever a graph has a tablelike flavor (Cox 2012).

```
. graph dot (mean) mpg, over(rep78) exclude0 l1title("`: var label rep78´")
> ysc(alt titlegap(*5)) ytitle(Mean miles per gallon)
```

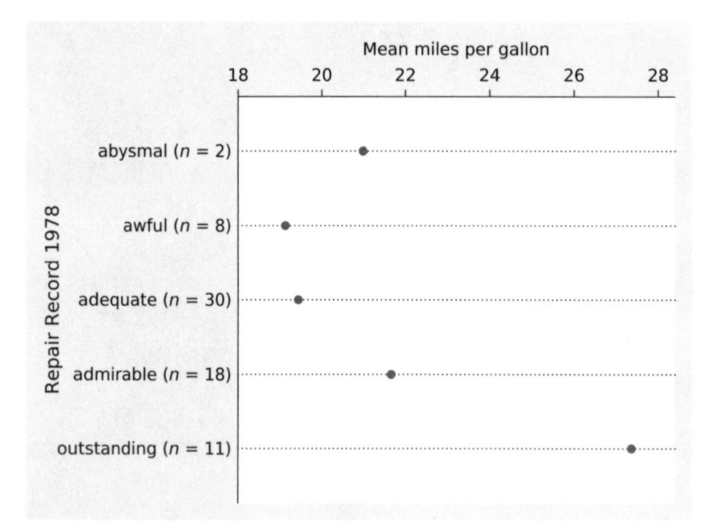

Figure 4. Dot chart for mean mpg by repair record, with whimsical value labels and a reminder of the number of observations.

You might prefer to do that interactively, say, by using `tabulate rep78` to get the frequencies and then typing in revised text in the Variables Manager `varm`. As before, the syntax is simple because there are no missing values of `mpg`.

References

Cleveland, W. S. 1994. *The Elements of Graphing Data*. Rev. ed. Summit, NJ: Hobart.

Cox, N. J. 2002. Speaking Stata: How to move step by: step. *Stata Journal* 2: 86–102. https://doi.org/10.1177/1536867X0200200106.

———. 2011. Stata tip 104: Added text and title options. *Stata Journal* 11: 632–633. https://doi.org/10.1177/1536867X1201100410.

———. 2012. Speaking Stata: Axis practice, or what goes where on a graph. *Stata Journal* 12: 549–561. https://doi.org/10.1177/1536867X1201200314.

———. 2020a. Speaking Stata: Loops, again and again. *Stata Journal* 20: 999–1015. https://doi.org/10.1177/1536867X20976340.

———. 2020b. Stata tip 139: The by() option of graph can work better than graph combine. *Stata Journal* 20: 1016–1027. https://doi.org/10.1177/1536867X20976341.

Wainer, H. 1984. How to display data badly. *American Statistician* 38: 137–147. https://doi.org/10.1080/00031305.1984.10483186.

———. 1997. *Visual Revelations: Graphical Tales of Fate and Deception from Napoleon Bonaparte to Ross Perot*. New York: Copernicus.

Wilke, C. O. 2019. *Fundamentals of Data Visualization: A Primer on Making Informative and Compelling Figures*. Sebastopol, CA: O'Reilly.

 DOI: 10.1177/1536867X221083928

The Stata Journal (2022)
22, Number 1, pp. 224–230

Stata tip 145: Numbering weeks within months

Nicholas J. Cox
Department of Geography
Durham University
Durham, U.K.
n.j.cox@durham.ac.uk

1 This day is in which week of its month?

A user wanted a variable to indicate in which week of its month a daily date fell. That question is a challenge both to imagine different definitions of weeks within months and to produce Stata code for each interpretation. It serves as a reminder that whatever date and time problems have been solved, there are still plenty more. See, for example, Cox (2010, 2012a,b, 2018b, 2019) for some previous notes in this territory.

Even if you lack interest in this specific question, it serves as an example for showing problem-solving skills in using Stata.

In what follows, we go no further than a standard Western calendar in which the months of each year are January to December. Much more on calendars can be found in standard references such as Blackburn and Holford-Strevens (1999) and Reingold and Dershowitz (2018). Unlike years, days, and months, weeks have no physical (meaning, astronomical) basis, but they do have many different associations and implications, ranging from mythological and religious to economic and cultural (for example, Henkin [2021]).

Surprisingly or not, Stata's built-in week functions are unlikely to be part of the answer. Stata's definition of a week is idiosyncratic. Week 1 always starts on 1 January, week 2 always starts on 8 January, and so on with 7-day weeks, until week 52 ends on 31 December and is 8 or 9 days long depending on whether the year is a leap year. There is no week 53 in this scheme. Stata's definition has one distinct advantage, which is that weeks always nest inside years without ever spanning two years. Otherwise, this definition seems rarely used outside Stata, and the rest of this tip is based on the assumption that you are using some different definition of a week.

We will work with a sandbox dataset with daily dates for the first three months of 2021. It is mental arithmetic to check that 2021 was not a leap year so that 90 = 31 + 28 + 31 observations will give us the right dataset size for those months.

```
. set obs 90
Number of observations (_N) was 0, now 90.
. generate ddate = mdy(12, 31, 2020) + _n
. format ddate %td
. list if inlist(_n, 1, _N)

     +-----------+
     |     ddate |
     |-----------|
  1. | 01jan2021 |
 90. | 31mar2021 |
     +-----------+

. generate mdate = mofd(ddate)
. format mdate %tm
```

For more on `_n` and `_N` if needed, see `help _variables`.

2 Daily and monthly dates introduced

Let me first recount some basics to help readers quite new to handling dates like these. If you are already broadly familiar with dates, you can skip and skim to the next section.

In Stata, dates are held as integers. Stata holds daily and monthly dates as integers, with the origin 0 as the first possible date in 1960. We have already used `mdy()` as a way to get daily dates from month, day, and year components. `daily("31 Dec 2020", "DMY")` is an example of one of several other ways to get started if daily dates arrive as strings. What is `mdy(12, 31, 2020)`? Let's use `display` to find out:

```
. display mdy(12, 31, 2020)
22280
```

It is 22280 on a scale in which 1 January 1960 is 0. Stata really does not expect mental arithmetic from you to work out in reverse that 22280 means 31 December 2020, which leads to the next point.

Date display formats show dates conventionally. It is the job of date display formats to show you what a date means conventionally. A format has nothing to do with what is stored—that remains here integers, like 22280—but, as said, only with what is displayed when you ask for it. Again, `display` shows this for small examples:

```
. display %td 22280
31dec2020
```

`%td` is the default daily date display format. Although a format can be used with `display` of individual values, it is more common, as shown earlier, to use the `format` command to associate variables with display formats.

Special date functions are key. I used the function `mofd()` (think monthly date of daily date) to go from daily dates to monthly dates, and then I applied `%tm` as the default monthly date format. To generate monthly date variables in Stata form directly, which is a different problem, I most often use one of two functions. `ym()` assembles a monthly date from numeric year and month components, just as `mdy()` assembles a daily date from numeric month, day, and year components. `monthly()` is for processing string arguments such as "Jan 2021", much as `daily()` does for daily dates in string form.

```
. display ym(2021, 1)
732
```

shows that, regardless of whatever display format is applied elsewhere, the monthly date for January 2021 is to Stata an integer, 732. Recall the principle that monthly dates too are stored as integers counting from 0, which is the first possible date in 1960, here January 1960.

Changing the date format does not change the date. It is often misunderstood, so let's underline that changing the date display format is never a way to convert from one kind of date to another (Cox 2012c). Short of setting up your own calculation, you need a conversion function such as `mofd()`.

We can now focus on the title problem.

3 Week 1 is days 1 to 7, and so on

The simplest definitions of each week of the month are that week 1 is days 1 to 7, week 2 days 8 to 14, and so on. On these definitions, there are usually 5 weeks in each month, and when that is true, the last week is 1, 2, or 3 days long depending on whether the month is 29, 30, or 31 days long. The exceptions are when the month is February in a nonleap year, so the number of days in the month is then 28, and there are 4 complete weeks.

Extracting day of month from a daily date is a common and fundamental need, so you should expect there to be a function provided to yield this directly. It is `day()`, which yields one of 1 to 31 as a result. We need to map days to week numbers, which could be

```
. generate wofm = ceil(day(ddate)/7)
```

That is it. From the inside outward, we read off `day(ddate)` and then divide by 7. So the result of that will be a decimal equivalent of 1/7, 2/7, 3/7, 4/7, 5/7, 6/7, 7/7, 8/7, and so on, up to 31/7. The `ceil()` function rounds up and is what we need: think "ceiling". It leaves integers unchanged and otherwise yields the next higher integer. That is, 1/7 to 7/7 round up to 1, 8/7 to 14/7 round up to 2, and so forth. `ceil()` and its sibling function `floor()` are often overlooked. I seize every opportunity to riff about them (Cox 2003, 2011, 2018a).

You might want to take that more slowly. You could put the result of `day(ddate)` in a new variable first:

```
. generate day = day(ddate)
```

Then, you could divide and round up:

```
. generate wofm = ceil(day/7)
```

Indeed, the last operation could also be split into two steps. You should certainly rewrite this way if you need a separate variable holding day of the month. It is also a good idea if the code is thereby clearer to you or anyone else needing to understand it.

It is prudent to check results. A programmer can rarely be too careful, but looking at a table is easy.

```
. tabulate mdate wofm
```

mdate	wofm 1	2	3	4	5	Total
2021m1	7	7	7	7	3	31
2021m2	7	7	7	7	0	28
2021m3	7	7	7	7	3	31
Total	21	21	21	21	6	90

That is as it should be. There are 4 complete weeks and 3 days left over for January and March (31 days), and there are 4 complete weeks for February (28 days in a nonleap year). An even more careful check would be to see whether the values 1 to 5 occur when they should and whether a leap year example works too.

4 A week starts on a Sunday, and so on

Another class of definitions might include a week starting on Sunday and ending on Saturday or starting on Monday and ending on Sunday. Do not rule out a weird-seeming definition that some group uses somewhere. My own university for many years had a convention that teaching weeks ran Thursday to Wednesday in the first term of the academic year and Monday to Friday in the other terms. The offset was to accommodate welcomes and whatnot at the beginning of each academic year. Admitting different start days may seem dismaying: do we need code for seven cases? Fortunately, one trick covers all.

Stata has a day-of-week function, `dow()`, that is handy here. `dow()` returns 0 for Sundays, 1 for Mondays, and so on, through to 6 for Saturdays. Let's take the case of starting on Sundays.

```
. bysort mdate (ddate): generate WOFM = sum(dow(ddate) == 0)
```

That is the essence. Ensuring that observations are in the right order, we use `bysort:` to calculate separately within each month. The code that follows the colon counts the number of Sundays so far in the month. `dow(date)` is equal to 0 when it is Sunday and not equal to 0 (because it is an integer from 1 to 6) on other days. When `dow(ddate) == 0` is true, then 1 is returned; when it is false, then 0 is returned. `sum()` is a key function that returns the running or cumulative sum of those 1s and 0s. The equivalent function in Mata is called `runningsum()`. In some statistical contexts, cumulative sums are called “cusums”, a term with some advantages and some disadvantages; either way, Stata does not use it generally, but note a distinct `cusum` command with a different goal.

If you are fuzzy about what `bysort:` does precisely, start with its help file or Cox (2002).

This is perhaps the trickiest point of the entire tip, so let us follow through what happens. At the start of each month, `dow(ddate) == 0` will be 1 or 0, and the running sum will be the same. As we look at each successive day in the same month, `dow(ddate) == 0` will alternate between 1s on Sundays and 0s on any other day. Correspondingly, the running sum will be bumped up by 1 on any Sunday and by 0 on any other day, meaning that the running sum will stay the same.

How does that pan out in the sandbox dataset? Again, a cross-tabulation is simple.

```
. tabulate mdate WOFM
```

mdate	WOFM 0	1	2	3	4	Total
2021m1	2	7	7	7	7	31
2021m2	6	7	7	7	1	28
2021m3	6	7	7	7	4	31
Total	14	21	21	21	12	90

mdate	WOFM 5	Total
2021m1	1	31
2021m2	0	28
2021m3	0	31
Total	1	90

In 2021, the first Sundays in each month were on 3 January, 7 February, and 7 March. Correspondingly, each month started with an incomplete week, which is coded 0, zero values being recorded because no Sundays had been observed so far. As it happens, August 2021 started on a Sunday, so no week 0 would be defined then.

Now twists on the idea should seem easy.

Starting on Monday or any other day of the week just means testing for a different result from `dow()`. If weeks start on Mondays, we are testing whether `dow(ddate) == 1`. So all seven cases for different starts and stops to the week yield to essentially the same trick.

Wanting the numbering of weeks to start at 1 just means adding 1 in calculating `WOFM` whenever the count starts at 0. Indeed, although starting the count at 0 may seem natural to anyone who knows more than a little mathematics, in the outside world of (say) medical or business reporting, it is much more likely that numbering starts at 1.

```
. by mdate: replace WOFM = WOFM + 1 if WOFM[1] == 0
(3 real changes made)
```

Notice the conditioning there. We want to bump up the counter if and only if the counter starts at 0.

5 Small morals

Given the extraordinary range of calendars and calendar practices across countries and subject areas, there could be yet other definitions. Yet we stop there and close with some small but standard morals to be drawn from the tale.

Dates are complicated. The entire rigmarole of dates (to say nothing of times too, but those are not an issue here) is a complicated mess. From your early education onward, you have been exposed repeatedly to the rules you usually need to know, but handling dates in software like Stata poses a new set of challenges. Some software handles dates through one or more distinct data or variable types. Stata's decision that dates and times are just integers simplifies much, but it still implies details about formats and functions that may need mastering. `help datetime` remains the best place to start.

Functions are helpful. The key to success here is knowing your functions, whether they are date functions that are what you need, such as `day()` or `dow()`, or other more general functions that help, such as `sum()` or `ceil()`. Functions remain the unsung heroes of Stata (Cox 2011).

Sandboxes are useful. Often, I witness people who have read large and complicated datasets into memory and are struggling to find the syntax they need. While that is the immediate problem, it can be a good idea to back up and create a small sandbox dataset, as we did here. You can get results quickly and compare easily with whatever you know to be the right answer. Other neglected ideas include using `display` to work with individual values and Mata as a glorified calculator (examples in Cox [2018a, 2019]).

References

Blackburn, B., and L. Holford-Strevens. 1999. *The Oxford Companion to the Year.* Oxford: Oxford University Press.

Cox, N. J. 2002. Speaking Stata: How to move step by: step. *Stata Journal* 2: 86–102. https://doi.org/10.1177/1536867X0200200106.

———. 2003. Stata tip 2: Building with floors and ceilings. *Stata Journal* 3: 446–447. https://doi.org/10.1177/1536867X0400300413.

———. 2010. Stata tip 68: Week assumptions. *Stata Journal* 10: 682–685. https://doi.org/10.1177/1536867X1101000409.

———. 2011. Speaking Stata: Fun and fluency with functions. *Stata Journal* 11: 460–471. https://doi.org/10.1177/1536867X1101100308.

———. 2012a. Stata tip 111: More on working with weeks. *Stata Journal* 12: 565–569. https://doi.org/10.1177/1536867X1201200316.

———. 2012b. Stata tip 111: More on working with weeks, erratum. *Stata Journal* 12: 765. https://doi.org/10.1177/1536867X1201200416.

———. 2012c. Stata tip 113: Changing a variable's format: What it does and does not mean. *Stata Journal* 12: 761–764. https://doi.org/10.1177/1536867X1201200415.

———. 2018a. Speaking Stata: From rounding to binning. *Stata Journal* 18: 741–754. https://doi.org/10.1177/1536867X1801800311.

———. 2018b. Stata tip 130: 106610 and all that: Date variables that need to be fixed. *Stata Journal* 18: 755–757. https://doi.org/10.1177/1536867X1801800312.

———. 2019. Speaking Stata: The last day of the month. *Stata Journal* 19: 719–728. https://doi.org/10.1177/1536867X19874247.

Henkin, D. M. 2021. *The Week: A History of the Unnatural Rhythms That Made Us Who We Are*. New Haven, CT: Yale University Press.

Reingold, N. M., and N. Dershowitz. 2018. *Calendrical Calculations: The Ultimate Edition*. Cambridge: Cambridge University Press.

The Stata Journal (2022)
22, Number 2, pp. 460–464 DOI: 10.1177/1536867X221106437

Stata tip 146: Using margins after a Poisson regression model to estimate the number of events prevented by an intervention

Milena Falcaro
King's College London
London, U.K.
milena.falcaro@kcl.ac.uk

Roger B. Newson
King's College London
London, U.K.
roger.newson@kcl.ac.uk

Peter Sasieni
King's College London
London, U.K.
peter.sasieni@kcl.ac.uk

After fitting a Poisson regression model to evaluate the effect of an intervention in a cohort study, one might be interested in estimating the number of events prevented by the intervention (assuming the observed associations are causal). This can be derived as the difference in the intervention group between the predicted number of events under the counterfactual (no intervention) and the factual (intervention) scenarios. One could use the `predict` command to obtain the predicted number of events under the two scenarios and then sum up the differences, but this approach would not be convenient for several reasons. One would need to change the intervention variable to get the counterfactual predicted values, and the confidence intervals would not be readily available (`bootstrap` or `jackknife` could be used, but this could be particularly time consuming if the dataset is large).

We here suggest using the `margins` command. Its use, however, is not straightforward for our specific problem because `margins` computes predictions for each observation (like `predict`) and then takes the average of these predicted values. For example, if our data are aggregated in years, `margins` will provide an average of the year-specific predictions. When `margins` is applied over N records and $\widehat{P}_i$ is the predicted value for the ith observation ($i = 1, \ldots, N$), the result is simply the average of these predicted values, that is, $\left(\sum_{i=1}^{N} \widehat{P}_i\right)/N$. If we want `margins` to calculate the sum of the predictions instead of the mean, we can multiply each observation-specific prediction by the number of observations (that is, N), and the result of `margins` will be $\left(\sum_{i=1}^{N} N\widehat{P}_i\right)/N = \sum_{i=1}^{N} \widehat{P}_i$.

Let's consider a simple example using simulated data. Specifically, we use a Poisson distribution to generate a variable, `cases`, containing the number of events of interest (for example, the number of cancer cases) as a function of an intervention indicator (`trt` = `1` if treated, `0` otherwise); two covariates (`x1` and `x2`); and an offset (`pyar` = person-years at risk).

```
. clear
. set seed 12345
. set obs 1000
. generate x1=runiform(50,100)
. generate x2=rbinomial(1,0.3)
. generate trt=rbinomial(1,0.5)
. generate pyar=runiformint(200,400)
. generate m=exp(0.01-0.2*trt-0.05*x1+0.8*x2 + ln(pyar))
. generate cases=rpoisson(m)
```

We then fit a Poisson regression model:

```
. poisson cases i.trt x1 i.x2, exp(pyar)
Iteration 0:   log likelihood = -2452.9776
Iteration 1:   log likelihood = -2452.9125
Iteration 2:   log likelihood = -2452.9125
Poisson regression                              Number of obs =     1,000
                                                LR chi2(3)    =   6654.25
                                                Prob > chi2   =    0.0000
Log likelihood = -2452.9125                     Pseudo R2     =    0.5756

------------------------------------------------------------------------------
       cases | Coefficient  Std. err.      z    P>|z|     [95% conf. interval]
-------------+----------------------------------------------------------------
       1.trt |  -.1925239   .0188441   -10.22   0.000    -.2294576   -.1555902
          x1 |  -.0497789    .000752   -66.19   0.000    -.0512529    -.048305
        1.x2 |   .7863367   .0188159    41.79   0.000     .7494582    .8232151
       _cons |   .0105769   .0514323     0.21   0.837    -.0902284    .1113823
   ln(pyar) |          1  (exposure)
------------------------------------------------------------------------------
```

To obtain an estimate of the number of events prevented by the intervention and its 95% confidence interval, the `margins` command will need to include the following (see [R] **margins** for more details):

- an `if` qualifier (that is, `if trt==1`) or the corresponding `subpop()` option (the latter must be used if the `vce(unconditional)` option is specified too);
- two `at()` options: one for the factual scenario (that is, `at((asobserved) _all)`) and one for the counterfactual scenario (that is, `at(trt=0)`);
- `expression(predict(n)*r)`, where `r` is the size of the group of observations over which the `margins` command averages the predictions (it is here retrieved from the two command lines `count if trt==1 & e(sample)==1` and `scalar r=r(N)`); and
- the `pwcompare` option.

Hence, the commands and results are as follows:

```
. count if trt==1 & e(sample)==1
  504

. scalar r=r(N)

. margins, at((asobs) _all) at(trt=0) exp(predict(n)*r) subpop(if trt==1)
> pwcompare

Pairwise comparisons of predictive margins          Number of obs    = 1,000
Model VCE: OIM                                      Subpop. no. obs =    504

Expression: predict(n)*r
1._at: (asobserved)
2._at: trt     = 0
```

	Contrast	Delta-method std. err.	Unadjusted [95% conf.	interval]
_at				
2 vs 1	1082.121	105.661	875.0296	1289.213

The above results show that the intervention is estimated to have prevented 1,082 (95% CI: 875 to 1,289) cancer cases in our sample. Had we used the above **margins** command without the **expression()** option, we would have obtained the average of the observation-specific predicted number of events:

```
. margins, at((asobs) _all) at(trt=0) subpop(if trt==1) pwcompare

Pairwise comparisons of predictive margins          Number of obs    = 1,000
Model VCE: OIM                                      Subpop. no. obs =    504

Expression: Predicted number of events, predict()
1._at: (asobserved)
2._at: trt     = 0
```

	Contrast	Delta-method std. err.	Unadjusted [95% conf.	interval]
_at				
2 vs 1	2.147066	.2096448	1.73617	2.557962

To better understand the above output, one can generate the variables (here called **pred1** and **pred2**) containing the observation-specific predictions for the two scenarios and then look at their means. The **pwcompare** option will be omitted because it is not allowed when the **generate()** option is specified too.

```
. margins, at((asobs) _all) at(trt=0) subpop(if trt==1) generate(pred)
Predictive margins                              Number of obs    = 1,000
Model VCE: OIM                                  Subpop. no. obs =    504

Expression: Predicted number of events, predict()
1._at: (asobserved)
2._at: trt    = 0

                          Delta-method
                  Margin   std. err.      z    P>|z|     [95% conf. interval]

         _at
          1      10.1131   .1416533    71.39   0.000      9.83546    10.39073
          2     12.26016   .1545487    79.33   0.000     11.95725    12.56307

. sum pred1 pred2
    Variable        Obs        Mean    Std. dev.       Min        Max

       pred1        504     10.1131    8.375062   1.288584   45.00473
       pred2        504    12.26016    10.15313   1.562158   54.55948
```

If we calculate the difference between the means of `pred2` (counterfactual scenario) and `pred1` (factual scenario), we obtain the value reported in the above `margins` command, where we omitted both the `expression()` and `pwcompare` options ($12.26016 - 10.1131 = 2.14706$). If we now generate the difference between `pred2` and `pred1` (that is, `generate diff=pred2-pred1`) and use the `total` command, we will obtain the point estimate reported by `margins` with the `expression()` and `pwcompare` options.

```
. generate diff=pred2-pred1
(496 missing values generated)

. total pred1 pred2 diff
Total estimation                        Number of obs = 504

                   Total   Std. err.     [95% conf. interval]

       pred1        5097   188.0197      4727.599    5466.401
       pred2    6179.121   227.9373      5731.295    6626.948
        diff    1082.121   39.91762      1003.696    1160.547
```

Extensions to interventions with two or more levels (for example, 0 = no treatment, 1 = low-dosage treatment, 2 = high-dosage treatment) or other counterfactual scenarios would be straightforward. For example, if we want to estimate how many fewer cases we would have observed in the nonintervention group (that is, `trt = 0`) if everybody had received the treatment, then we would specify the following:

```
. quietly poisson cases i.trt x1 i.x2, exp(pyar)

. count if trt==0 & e(sample)==1
  496

. scalar s=r(N)

. margins, at((asobs) _all) at(trt=1) exp(predict(n)*s) subpop(if trt==0)
> pwcompare

Pairwise comparisons of predictive margins             Number of obs    = 1,000
Model VCE: OIM                                         Subpop. no. obs =    496

Expression: predict(n)*s
1._at: (asobserved)
2._at: trt    = 1

─────────────┬──────────────────────────────────────────────────
             │                 Delta-method          Unadjusted
             │    Contrast   std. err.     [95% conf. interval]
─────────────┼──────────────────────────────────────────────────
         _at │
      2 vs 1 │   -1106.267   107.9831      -1317.91   -894.6243
─────────────┴──────────────────────────────────────────────────
```

Thus, our model estimates that if everyone in the nonintervention group had been administered the treatment, there would have been 1,106 (95% CI: 895 to 1318) fewer cancer cases. Note that the contrast is negative, corresponding to fewer cases had everyone been treated. This is because we are comparing the counterfactual scenario represented by `at(trt=1)` (that is, scenario 2 = untreated patients are treated) versus the factual scenario specified by `at((asobs) _all)` (that is, scenario 1 = untreated patients are untreated).

What is discussed in this Stata tip could also be extended to case–control studies by using inverse-probability-of-sampling weights to estimate absolute rates.

Acknowledgment

This work was supported by Cancer Research U.K. (grant number: C8162/A27047).

146 DOI: 10.1177/1536867X221106438

The Stata Journal (2022)
22, Number 2, pp. 465–466

Erratum: Stata tip 145: Numbering weeks within months

Nicholas J. Cox
Department of Geography
Durham University
Durham, U.K.
n.j.cox@durham.ac.uk

The last line of code in Cox (2022) is quite wrong. To explain the problem, let me back up and give you enough context to understand what was written and why it is wrong.

The tip uses a sandbox dataset with daily and monthly dates for the first three months of 2021.

```
. clear
. set obs 90
. generate ddate = mdy(12, 31, 2020) + _n
. format ddate %td
. generate mdate = mofd(ddate)
. format mdate %tm
```

The goal is—given an example definition that weeks start on Sundays, for which `dow(ddate)` returns 0—to number weeks within months. A first stab at this numbering is

```
. bysort mdate (ddate): generate WOFM = sum(dow(ddate) == 0)
```

The simplest case is that a month starts on a Sunday. If so, that line of code yields 1 for the first 7 days of a month (by this definition, the first week), 2 for the next 7 days, and so on. There will usually be an incomplete week 5. The exception is any February in a nonleap year that starts on a Sunday. But in all other cases, in which a month starts on a day other than Sunday, the week numbering starts with 0. Suppose this is not what you want, because you prefer week numbering to start at 1. The tip offers this fix:

```
. by mdate: replace WOFM = WOFM + 1 if WOFM[1] == 0
```

Perhaps, you can see already why this is wrong, but here is the story. If the value of `WOFM` for the first day in a month is indeed 0, then it is changed to 1. But now it is no longer true that `WOFM[1]` is 0 by virtue of the change just made. So no further changes can take for observations in the same month. Hence, the fix is correct only when the first day of the month is a Saturday and only one value needs to be changed.

The easiest fix is to `generate` a new variable.

```
. by mdate: gen WOFM2 = WOFM + 1 if WOFM[1] == 0
```

A similar fix would apply to weeks starting on any other day of the week. The statement generating `WOFM` would need a variant on `dow(ddate) == 0`, that is, the appropriate value between 1 and 6 rather than 0.

Many thanks to Leonardo Guizzetti for flagging this error.

Reference

Cox, N. J. 2022. Stata tip 145: Numbering weeks within months. *Stata Journal* 22: 224–230. https://doi.org/10.1177/1536867X221083928.

DOI: 10.1177/1536867X221141067
The Stata Journal (2022)
22, Number 4, pp. 996–997

Stata tip 147: Porting downloaded packages between machines

Roger B. Newson
Cancer Prevention Group
King's College London
London, U.K.
roger.newson@kcl.ac.uk

Stata users are often asked to work on web-disabled machines with little or no Internet access. So it may not be possible for the user to download packages at will using the `net` or `ssc` command. This requirement may exist for good reasons, such as prevention of cybercrime and data breaches.

Stata, like R and Genstat, is a statistical language with potentially as many dialects as users. These dialects are defined by the optional packages that each user has downloaded; the number may be hundreds for a high-power user. This is a major advantage of a statistical language (such as Stata) over a statistical package (such as SPSS). And Stata users would like to continue to have that advantage on web-disabled machines.

Fortunately, downloaded packages can be ported between machines. Downloaded packages on a web-enabled machine are stored in a folder called `PLUS`, which is one of a list of folders called the ado-path and which Stata searches when asked to run an ado-file. To list the ado-path, including the `PLUS` folder, we should use the **`adopath`** command (see [P] **sysdir**). If we use the author's Stata 17 under Windows 10, we get the following output:

```
. adopath
  [1]  (BASE)      "C:\Program Files\Stata17\ado\base/"
  [2]  (SITE)      "C:\Program Files\Stata17\ado\site/"
  [3]              "."
  [4]  (PERSONAL)  "C:\Users\Roger Newson\ado\personal/"
  [5]  (PLUS)      "C:\Users\Roger Newson\ado\plus/"
  [6]  (OLDPLACE)  "c:\ado/"
```

We see that the `PLUS` folder is

```
"C:\Users\Roger Newson\ado\plus/"
```

which is a folder containing subfolders that contain the files belonging to the downloaded packages. In our case, there are over 100 packages containing over 1,000 files.

Most ways of porting a `PLUS` folder start by zipping it into a `.zip` folder to save space. The official Stata way to do this is by using the **`zipfile`** command (see [D] **zipfile**):

```
. zipfile "C:\Users\Roger Newson\ado\plus/", saving(myplus.zip, replace)
```

This command produces a lot of output to the log, which I have omitted. However, it also produces the `.zip` folder `myplus.zip`, which contains the folder `Users`, which contains the folder `Roger Newson`, which contains the folder `ado`, which contains the folder `plus`, which contains the subfolders and other contents of our original `PLUS` folder.

A `.zip` folder may be ported using a high-security file transfer utility, or it may be simply copied to a USB drive that is itself ported to the destination machine. Either way, the user copies the contents of the ported `PLUS` folder to a location on the destination machine. In our case, our destination machine also uses Windows, and this location will be

```
S:\CPG\Studies\cpgplus\
```

To be able to use the packages on our destination machine, we simply add the location (and its subfolders) to our ado-path, using an `adopath +` command:

```
. adopath + "S:\CPG\Studies\cpgplus\"
  [1]                "."
  [2]  (BASE)        "C:\Program Files\Stata17\ado\base/"
  [3]  (SITE)        "C:\Program Files\Stata17\ado\site/"
  [4]  (PERSONAL)    "D:\Users\Roger.Newson\ado\personal/"
  [5]  (PLUS)        "D:\Users\Roger.Newson\ado\plus/"
  [6]  (OLDPLACE)    "c:\ado/"
  [7]                "S:\CPG\Studies\cpgplus\"
```

Once this is done, we can use any packages in the ported folder (or its subfolders). Note that the ported folder is ported as a single item, so this process takes no more time for 100 packages containing 1,000 files than for 1 package containing 2 files. Therefore, no powers are lost, even for high-power users with hundreds of downloaded packages.

Methods similar to these might also save time porting all of a user's downloaded packages together from one machine to another, even if the destination machine is not web disabled or if the downloaded packages on the source machine live in a folder other than the `PLUS` folder.

150 DOI: 10.1177/1536867X221141068

The Stata Journal (2022)
22, Number 4, pp. 998–1003

Stata tip 148: Searching for words within strings

Nicholas J. Cox
Department of Geography
Durham University
Durham, U.K.
n.j.cox@durham.ac.uk

1 The problem: Looking for words

Searching for particular text within strings is a common data management problem. One frequent context is whenever various possible answers to a question are bundled together in values of a string variable. Suppose people are asked which sports they enjoy or something more interesting, like which statistical software they use routinely. To keep the matter simple, we will first imagine just lists of one or more numbers that are concise codes for distinct answers, say, `"42"` for `"cricket"` or `"1"` for `"Stata"`. Nonnumeric codes will also be considered in due course. For more on handling such questions, sometimes called multiple response, see Cox and Kohler (2003) or Jann (2005).

The precise problem discussed in this tip is finding text in strings whenever such text is a word in Stata's sense, or something close to that. This needs a little explanation.

Here is a tiny sandbox dataset that will be enough to show the problem and some devices that can yield solutions. By way of example, we will focus mainly on a goal of generating indicator variables, sometimes known as dummy variables. For one overview of generating such variables, see Cox and Schechter (2019). We will also touch on the problem of counting instances of a word.

```
. input str8 mytext
        mytext
  1. "1"
  2. "1 2"
  3. "1 2 11"
  4. "11 12 13"
  5. "11 12 13 111"
  6. end
```

Searching for `"1"` or `"2"`, say, starts with looking for either character with a string function. The function `strpos()` is useful for that. For a rapid personal survey of especially useful functions, see Cox (2011a).

Finding such single characters is easy and unproblematic if the possible answers are one character long at most. More generally, searches are easy if there is no ambiguity. Consider

```
. generate byte is1 = strpos(mytext, "1") > 0
```

The function `strpos()` looks for particular text within other text. It returns 0 if that particular text is not found and a positive number, the position of that particular text, if that text is found. Thus, the position of `"1"` in `"1 2"` is 1, the position of `"2"` in

"1 2" is 3, and so on. Hence, an indicator variable like is1 will be returned as 1 if there are observations in which strpos() returns a positive result. Otherwise, the indicator variable will be returned as 0. If you are new to the idea that an expression like

```
strpos(mytext, "1") > 0
```

returns 1 if true and 0 if false, see Cox and Schechter (2019) or, more directly, Cox (2005, 2016).

If you look again at the sandbox, you should see what is coming next. Looking for "1" with

```
strpos("1 2 11", "1")
```

will still work, fortunately, but looking for "1" with

```
strpos("11 12 13", "1")
```

will yield a false positive. The problem is that we want to find "1" only if it occurs by itself, namely, as a separate word. Stata's primary sense of a word within a string is that words are separated by spaces.

In some Stata contexts, double quotation marks bind together more strongly than spaces separate, so "Stata is subtle" would be treated as a single word if the quotation marks were explicit. For present purposes, we will leave that complication aside.

2 A solution: Looking for spaces too

Let's carry forward the idea that we need to look for spaces too. At first sight, this is a beautiful idea that just does not work very well because there are too many possibilities to catch. Thus, looking for "1 " catches "1"—as part of "1 "—and not "11" within "1 2 11", which is as intended. But it catches the first "1 "—as part of "11 "—within "11 12 13", which is not what we want. Other way round, looking for " 1" catches correctly sometimes and incorrectly other times. Looking for " 1 "—with spaces before and after—will not work if "1" is the first word or the last word, so without a previous space or a following space, respectively.

But that last idea can be made to work with a simple twist. Congratulations if you thought of this directly!

```
. generate byte is1 = strpos(" " + mytext + " ", " 1 ")
. list
```

	mytext	is1
1.	1	1
2.	1 2	1
3.	1 2 11	1
4.	11 12 13	0
5.	11 12 13	0

So we solve the problem of initial and following spaces by supplying them on the fly. Note that we do not need to `generate` a new variable or `replace` an existing variable; we just get Stata to work with a version of the variable with extra spaces. Extra spaces that go beyond our need are harmless, because `"  1  "`, in which "1" has two spaces before it and two after it, is treated the same way as `" 1 "`, in which "1" has one space before it and one after it.

3 What about other separators?

Suppose our string variable used another separator, say, commas, which could just be a different convention or a good idea anyway if spaces occur naturally. Someone's favorite sport might be `"water polo"` or `"debugging code"`. Then whatever the commas separate are not words in Stata's technical sense, but they are still words for us or atoms we wish to seek as such.

We could still use a similar idea of looking for `",1,"` within `"," + mytext + ","`. We just need to watch for gratuitous extra spaces so that `"1 ,"` is not missed. If strings could be moderately complicated, we might need a different method. More positively, if spaces have no meaning and we have values like `"1,2 ,3"`, then changing all commas to spaces allows the method of the previous section to be used.

4 A solution: What would change if we deleted words?

Here is another solution. This time around, an example comes before the explanation.

```
. generate byte IS1 = strlen(mytext) > strlen(subinword(mytext, "1", "", 1))
. list
```

	mytext	is1	IS1
1.	1	1	1
2.	1 2	1	1
3.	1 2 11	1	1
4.	11 12 13	0	0
5.	11 12 13	0	0

We get the same answer, so how did that work?

The function `strlen()` measures the length of strings by counting characters. Although no longer documented, the older name `length()` still works if you remember or prefer that.

The function `subinword()` replaces text with other text if and only if that text occurs as a word in Stata's primary sense. The function knows how to handle words at the beginning and end of strings. However, `subinword()` does not follow Stata's extended sense that a word can be defined (meaning, delimited) by explicit double quotation marks.

But how does replacing text help? We do not want to change text; we are just searching for it. Yet, if the result of replacing text by an empty string (deleting it, to put it plainly) would be to reduce the length of the string, then evidently we did find that text.

Notice "would be". As before, we do not have to `generate` a new variable or `replace` an existing variable. We just get Stata to tell us what the result would be if the text existed and so would be deleted.

Whether the length of the string is greater than the length of the string with the word removed is a true or false question. Either the first length is greater because there is at least one instance of the word or the two lengths are the same because there is no such instance. If the expression is true, 1 is returned; and if it is false, 0 is returned, giving us an indicator variable.

This method is of interest for another reason: you may want to count instances of a word. We could have written

```
. generate byte IS1 = strlen(mytext) > strlen(subinword(mytext, "1", "", .))
```

The difference is in the last argument fed to `subinword()`, namely, system missing . rather than 1. That different syntax instructs Stata to delete all instances of the word `"1"` rather than the first only. For detecting whether the word exists, you need know only that it exists at least once.

If the problem is counting instances instead of checking for existence, then the difference in lengths

```
. generate count1 = strlen(mytext) - strlen(subinword(mytext, "1", "", .))
```

is precisely the number of times `"1"` occurs as a word. If you are looking for instances of `"11"` or `"111"`, remember to divide by 2 or 3—the lengths of the words in question, respectively—or you will get the number of characters notionally deleted, not the number of words.

For more on counting substrings, see Cox (2011b).

5 Nonnumeric words

Datasets may include one or more nonnumeric words bundled in a string variable. Suppose there was a survey question about which programming languages are routine for Stata users, with possible answers such as one or more of `Python`, `Julia`, `C++`, and `C`.

Handling such nonnumeric words can be both easier and more difficult than handling numeric words. The possibility of ambiguity is less but still present, as witness checking for mentions of `C` and finding them within mentions of `C++`. Hence, insisting on searching for a word, and not just a substring, can be necessary using one of the devices just explained.

Greater difficulty can arise because of variations in spelling and punctuation, depending sensitively on how such data were entered and collated. Suppose that `none` was expected as an answer when true but that there are also instances of `None`, `NONE`, and so forth. This particular variability is easily handled by looking for `none` within `strlower()` or—according to taste—looking for `NONE` within `strupper()`. The older function names `lower()` and `upper()` are equivalent and still work. Other variations in spelling may be harder to handle, but the first step is always to find out exactly which names were used.

6 A list of tricks

We have covered two main ideas:

- Words are separated by spaces, so look for a word together with previous and following spaces, remembering how to catch words at the beginning or the end of a string (sections 2 and 3).
- If we ask Stata to tell us whether and how the length of a string would change if we were to delete a word, we have ways to detect the occurrence of that word, either yes or no, or the number of occurrences if that is what we seek (section 4).

That is not a complete treatise, even on this small topic. A longer account might mention other possibilities, complications that may arise, or possible solutions.

First, I will mention other problems:

- I have focused on plain ASCII characters, but searching for Unicode needs more care and different functions.
- I have mentioned but not fully solved the complication of "words" that include spaces. But the more complicated the string we are searching for, the less likely ambiguity is to bite.
- I have focused on simple searching of string variables, but string manipulation is needed in other contexts, such as parsing user input if you are writing Stata programs.

Now, I will signal other solutions:

- Many readers will already know about regular expression syntax.
- Sometimes, we cannot solve a problem with one command line. We may need to use the `gettoken` (see [P] **gettoken**) command or the `split` (see [D] **split**) command. We may need to loop over words with a construct like `foreach` or `forvalues` (see [P] **foreach** or [P] **forvalues**).

All of these matters deserve detailed treatment, which is left to other accounts.

7 Acknowledgment

William Lisowski made helpful comments on a draft.

References

Cox, N. J. 2005. FAQ: What is true or false in Stata? https://www.stata.com/support/faqs/data-management/true-and-false/.

———. 2011a. Speaking Stata: Fun and fluency with functions. *Stata Journal* 11: 460–471. https://doi.org/10.1177/1536867X1101100308.

———. 2011b. Stata tip 98: Counting substrings within strings. *Stata Journal* 11: 318–320. https://doi.org/10.1177/1536867X1101100212.

———. 2016. Speaking Stata: Truth, falsity, indication, and negation. *Stata Journal* 16: 229–236. https://doi.org/10.1177/1536867X1601600117.

Cox, N. J., and U. Kohler. 2003. Speaking Stata: On structure and shape: The case of multiple responses. *Stata Journal* 3: 81–99. https://doi.org/10.1177/1536867X0300300106.

Cox, N. J., and C. B. Schechter. 2019. Speaking Stata: How best to generate indicator or dummy variables. *Stata Journal* 19: 246–259. https://doi.org/10.1177/1536867X19830921.

Jann, B. 2005. Tabulation of multiple responses. *Stata Journal* 5: 92–122. https://doi.org/10.1177/1536867X0500500113.

156 DOI: 10.1177/1536867X231162021

The Stata Journal (2023)
23, Number 1, pp. 276–280

Stata tip 149: Weighted estimation of fixed-effects and first-differences models

John Gardner
Department of Economics
University of Mississippi
University, MS
jrgardne@olemiss.edu

Applied econometricians frequently use weighted regressions to improve the precision of fitted panel-data models. For example, suppose that the outcome y_{igt} for individual i in group g at time t is

$$y_{igt} = \mathbf{x}'_{gt}\boldsymbol{\beta} + c_g + \varepsilon_{igt}$$

A group-average version of this model is

$$y_{gt} = \mathbf{x}'_{gt}\boldsymbol{\beta} + c_g + \varepsilon_{gt} \tag{1}$$

where $y_{gt} = \sum_i y_{igt}/n_{gt}$, n_{gt} is the number of observations in group g at time t and ε_{gt} is defined similarly. The group-average model might be relevant because individual-level data are not available (for example, because of confidentiality concerns) or for computational reasons.

In such cases, it is common practice to weight the model by n_{gt}.[1] The justification for this practice is that, if the original ε_{igt} are homoskedastic and serially uncorrelated with variance σ^2, then $V(\varepsilon_{gt}) = \sigma^2/n_{gt}$, and the Gauss–Markov theorem applies to the weighted model, which has homoskedastic errors.[2]

This tip clarifies estimation of weighted panel-data models in Stata in two ways. First, it extends the well-known deviation-from-means interpretation of fixed-effects models and the equivalence between fixed-effects and first-differences models with two time periods to the case of weighted estimation. Second, it highlights several ways to fit weighted fixed-effects (WFE) models in Stata. Of course, the tip also applies to models that are weighted for reasons other than heteroskedasticity arising from group averaging.

1. This can be accomplished in Stata using analytic weights, which are "inversely proportional to the variance of an observation" (StataCorp 2021). When you insert the analytic weight into the calculation formula, "you are treating each observation as one or more real observations" (StataCorp 2021). In the regression context, least-squares estimation weighted by n_{gt} is equivalent to least-squares estimation of a transformed model in which each variable for each observation is multiplied by $\sqrt{n_{gt}}$.
2. This is not necessarily a good idea. Solon, Haider, and Wooldridge (2015) show that, if the ε_{igt} are autocorrelated (for example, because of clustering), weighting may increase the estimated standard errors.

To illustrate weighted estimation of models such as (1) in Stata, I begin by generating some heteroskedastic panel data:

```
. set seed 57474

. set obs 100
Number of observations (_N) was 0, now 100.

. generate c = rnormal(1, 2)                          // fixed effects

. generate g = _n                                     // groups

. forvalues t=1/2 {
  2.        generate n`t´ = max(1,ceil(uniform()*100)) // group sample sizes
  3.        generate x`t´ = rnormal(c) + rnormal()     // x_gt correlated with c_g
  4.        generate e`t´ = rnormal(0, 5/sqrt(n`t´))   // heteroskedastic errors
  5.        generate y`t´ = 5 + 2*x`t´+ c +e`t´        // y_gt
  6. }

. reshape long x e y n, i(g) j(t)
(j = 1 2)

Data                               Wide   ->   Long
-----------------------------------------------------------------------------

Number of observations              100   ->   200
Number of variables                  10   ->   7
j variable (2 values)                     ->   t
xij variables:
                                  x1 x2   ->   x
                                  e1 e2   ->   e
                                  y1 y2   ->   y
                                  n1 n2   ->   n
-----------------------------------------------------------------------------
```

The simplest route to weighted estimation is via the `regress` command, with group dummy variables and analytic weights equal to the group-time sample sizes:

```
. regress y x i.g [aw=n]
(sum of wgt is 10,289)

      Source |       SS           df       MS      Number of obs   =       200
-------------+----------------------------------   F(100, 99)      =    156.02
       Model |  7952.52137       100  79.5252137   Prob > F        =    0.0000
    Residual |  50.4613647        99  .509710754   R-squared       =    0.9937
-------------+----------------------------------   Adj R-squared   =    0.9873
       Total |  8002.98274       199  40.2159937   Root MSE        =    .71394

------------------------------------------------------------------------------
           y | Coefficient  Std. err.      t    P>|t|     [95% conf. interval]
-------------+----------------------------------------------------------------
           x |   2.001036   .0557417    35.90   0.000     1.890432     2.11164
             |
           g |
           2 |  -1.184803   .7453728    -1.59   0.115    -2.663785    .2941781
           3 |   1.455865   .5802503     2.51   0.014     .3045227    2.607208
  (output omitted)
------------------------------------------------------------------------------
```

In this case, the weighted estimate compares favorably with the unweighted point estimate of 1.97 with standard error 0.075 (not shown).

In the unweighted case, the fixed-effects dummy-variable estimator has a deviation-from-means interpretation: it can be obtained by a *within* regression that replaces y_{gt}

and $\mathbf{x}_{gt}$ with the deviations of those variables from their group-specific means (eliminating the c_g along the way). A natural question is whether WFE estimation has a similar interpretation.

The Frisch–Waugh–Lovell theorem (see, for example, Greene [2018, theorem 3.2]) implies that WFE estimates can be obtained from a weighted regression that replaces y_{gt} and $\mathbf{x}_{gt}$ with the residuals from weighted regressions of those variables on a full set of group dummies. To connect this to a deviation from means interpretation, note that, because the group dummies are mutually orthogonal, the coefficient on the dummy d_{jgt} for group j from a weighted regression of y_{gt} on a full set of group dummies (and no overall constant) can be obtained from a weighted regression of y_{gt} on d_{jgt} alone as

$$\widehat{\lambda}_j = \frac{\sum_{g,t} n_{gt} y_{gt} d_{jgt}}{\sum_{g,t} n_{gt} d_{jgt}^2} = \frac{\sum_{g,t} n_{gt} y_{gt} d_{jgt}}{\sum_{g,t} n_{gt} d_{jgt}} = \frac{\sum_t n_{jt} y_{jt}}{\sum_t n_{jt}}$$

and similarly for $\mathbf{x}_{gt}$. Consequently, weighted dummy-variable estimation of (1) is equivalent to least-squares estimation of the weighted model

$$\sqrt{n_{gt}}\left(y_{gt} - \frac{\sum_t n_{gt} y_{gt}}{\sum_t n_{gt}}\right) = \sqrt{n_{gt}}\left(\mathbf{x}_{gt} - \frac{\sum_t n_{gt} \mathbf{x}_{gt}}{\sum_t n_{gt}}\right)' \boldsymbol{\beta} + \sqrt{n_{gt}}\left(\varepsilon_{gt} - \frac{\sum_t n_{gt} \varepsilon_{gt}}{\sum_t n_{gt}}\right) \tag{2}$$

In other words, the weighted dummy-variable estimator is equivalent to a *weighted* within estimator that replaces y_{gt} and $\mathbf{x}_{gt}$ with deviations from their *weighted* means. This estimator may be preferable when the number of groups is large.

The following illustrates this weighted-deviation-from-weighted-means interpretation:

```
. bysort g: egen sumn=sum(n)
. foreach z in x y {
  2.         generate `z´w=`z´*n
  3.         bysort g: egen `z´wsum = sum(`z´w) // weighted sums
  4.         generate `z´wbar = `z´wsum/sumn    // weighted means
  5.         generate `z´dev = `z´-`z´wbar      // deviations from weighted means
  6. }
. regress ydev xdev [aw=n], nocons
(sum of wgt is 10,289)

      Source |       SS           df       MS      Number of obs   =       200
-------------+----------------------------------   F(1, 199)       =   2590.40
       Model |  656.859929         1  656.859929   Prob > F        =    0.0000
    Residual |  50.4613654       199    .2535747   R-squared       =    0.9287
-------------+----------------------------------   Adj R-squared   =    0.9283
       Total |  707.321294       200  3.53660647   Root MSE        =    .50356

------------------------------------------------------------------------------
        ydev | Coefficient  Std. err.      t    P>|t|     [95% conf. interval]
-------------+----------------------------------------------------------------
        xdev |   2.001036   .0393162    50.90   0.000     1.923506    2.078566
------------------------------------------------------------------------------
```

Although the weighted-within point estimates are identical to the dummy-variable estimates, the standard errors are incorrect because they fail to account for the degrees of freedom used in computing the group-level weighted means.[3] Fortunately, the `areg` command does just that:

```
. areg y x [aw=n], absorb(g)
(sum of wgt is 10,289)

Linear regression, absorbing indicators         Number of obs     =        200
Absorbed variable: g                            No. of categories =        100
                                                F(1, 99)          =    1288.69
                                                Prob > F          =     0.0000
                                                R-squared         =     0.9937
                                                Adj R-squared     =     0.9873
                                                Root MSE          =     0.7139

------------------------------------------------------------------------------
           y | Coefficient  Std. err.      t    P>|t|     [95% conf. interval]
-------------+----------------------------------------------------------------
           x |   2.001036   .0557417    35.90   0.000     1.890432     2.11164
       _cons |   5.644418   .0697073    80.97   0.000     5.506104    5.782733
------------------------------------------------------------------------------
F test of absorbed indicators: F(99, 99) = 6.198              Prob > F = 0.000
```

The `xtreg` command with the `fe` option fits fixed-effects models similarly. However, because `xtreg` does not support time-varying weights, it cannot be used in this application.[4]

Another way to eliminate the group fixed effects in (1) is via the first-differences model

$$\Delta y_{gt} = \Delta \mathbf{x}'_{gt}\boldsymbol{\beta} + \Delta\varepsilon_{gt} \tag{3}$$

Empiricists frequently weight first-differenced models of group averages by $1/(1/n_{gt} + 1/n_{g,t-1})$, the justification being that, if the individual-level errors are homoskedastic and serially uncorrelated, then $V(\Delta\varepsilon_{gt}) = \sigma^2(1/n_{gt} + 1/n_{g,t-1})$.

In the unweighted case, it is well known that fixed effects and first differences are identical when there are only two time periods. Thus, it may not be surprising that fixed effects weighted by n_{gt} and first differences weighted by $1/(1/n_{gt} + 1/n_{g,t-1})$ are also identical in this case, as the following demonstrates:

```
. xtset g t

Panel variable: g (strongly balanced)
 Time variable: t, 1 to 2
         Delta: 1 unit

. generate wt=1/(1/n+1/l.n)
(100 missing values generated)
```

3. The correct degrees of freedom is $N(T-1) - K$, where N is the number of panel units, T is the number of time periods, and K is the number of regressors (this is also the degrees of freedom for a regression of y_{gt} on $\mathbf{x}_{gt}$ and a full set of N group dummies). If the ε_{gt} are independently distributed, valid standard errors can be obtained by multiplying the default standard errors by $\sqrt{(NT-K)/\{N(T-1)-K\}}$.
4. On the other hand, the **areg** (see [R] **areg**) command is not designed for applications where the number of groups increases with the sample size.

```
. regress d.y d.x [aw=wt], nocons
(sum of wgt is 2,114.55251610279)
      Source |       SS           df       MS      Number of obs   =       100
-------------+----------------------------------   F(1, 99)        =   1288.69
       Model |  1598.07611         1  1598.07611   Prob > F        =    0.0000
    Residual |  122.767572        99  1.24007648   R-squared       =    0.9287
-------------+----------------------------------   Adj R-squared   =    0.9279
       Total |  1720.84368       100  17.2084368   Root MSE        =    1.1136

------------------------------------------------------------------------------
         D.y | Coefficient  Std. err.      t    P>|t|     [95% conf. interval]
-------------+----------------------------------------------------------------
           x |
         D1. |   2.001036   .0557417    35.90   0.000     1.890432     2.11164
------------------------------------------------------------------------------
```

To see why this holds, note that, in the two-period case, the left-hand side of (2) is

$$\sqrt{n_{gt}}\frac{n_{gt'}(y_{gt} - y_{gt'})}{n_{gt} + n_{gt'}}$$

and similarly for the right-hand side. Thus, the WFE estimate of $\boldsymbol{\beta}$ is

$$\begin{aligned}
\widehat{\boldsymbol{\beta}}^{\mathrm{WFE}} &= \left\{\sum_{g,t} \frac{n_{gt}n_{gt'}^2(\mathbf{x}_{gt} - \mathbf{x}_{gt'})(\mathbf{x}_{gt} - \mathbf{x}_{gt'})'}{(n_{g1} + n_{g2})^2}\right\}^{-1} \\
&\quad \left\{\sum_{g,t} \frac{n_{gt}n_{gt'}^2(\mathbf{x}_{gt} - \mathbf{x}_{gt'})(y_{gt} - y_{gt'})}{(n_{g1} + n_{g2})^2}\right\} \\
&= \left\{\sum_{g} \frac{(n_{g1} + n_{g2})n_{g1}n_{g2}\Delta\mathbf{x}_{gt}\Delta\mathbf{x}_{gt}'}{(n_{g1} + n_{g2})^2}\right\}^{-1} \left\{\sum_{g} \frac{(n_{g1} + n_{g2})n_{g1}n_{g2}\Delta\mathbf{x}_{gt}\Delta y_{gt}}{(n_{g1} + n_{g2})^2}\right\} \\
&= \left(\sum_{g} \frac{n_{g1}n_{g2}\Delta\mathbf{x}_{gt}\Delta\mathbf{x}_{gt}'}{n_{g1} + n_{g2}}\right)^{-1} \left(\sum_{g} \frac{n_{g1}n_{g2}\Delta\mathbf{x}_{gt}\Delta y_{gt}}{n_{g1} + n_{g2}}\right)
\end{aligned}$$

The last expression is precisely the vector of coefficients $\widehat{\boldsymbol{\beta}}^{\mathrm{WFE}}$ on $\Delta\mathbf{x}_{gt}$ from a weighted least-squares estimate of (3).

References

Greene, W. H. 2018. *Econometric Analysis.* 8th ed. New York: Pearson.

Solon, G., S. J. Haider, and J. M. Wooldridge. 2015. What are we weighting for? *Journal of Human Resources* 50: 301–316. https://doi.org/10.3368/jhr.50.2.301.

StataCorp. 2021. *Stata 17 User's Guide.* College Station, TX: Stata Press.

The Stata Journal (2023)
23, Number 1, pp. 281–292 DOI: 10.1177/1536867X231162020

Stata tip 150: When is it appropriate to xtset a panel dataset with panelvar only?

Carlo Lazzaro
Studio di Economia Sanitaria
Milan, Italy
and
School of Pharmacology
Biology and Biotechnologies Department "Lazzaro Spallanzani"
University of Pavia
Pavia, Italy
carlo.lazzaro@tiscalinet.it

1 Introduction

The Stata command `xtset` (see [XT] **xtset**) is the requirement to access the `xt` suite of commands, which was developed to deal with datasets having both a cross-sectional (or N) and a time-series (or T) dimension (that is, panels) (Cameron and Trivedi 2005, 2022; Wooldridge 2020).

A panel dataset can be `xtset` in five ways. One of them allows the panel dataset to be `xtset` via the *panelvar* only:

Syntax 1

`xtset` *panelvar*

The code above tells Stata that the dataset is composed of panels, but the order of the observations belonging to each panel is irrelevant. The remaining four ways to `xtset` the panel dataset require a *timevar* too, with or without some additional options, to tell Stata how frequently observations are collected (for example, every two years):

Syntax 2

`xtset` *panelvar timevar* [, *tsoptions*]

2 Why does error r(451) occur?

From 2014, the Stata forum reports 500 queries (keyword: "repeated time values within panel"; last check July 6, 2022) concerning the error `r(451)`, the description of which can be accessed by typing the following from within Stata:

```
. search r(451)
[P]     error . . . . . . . . . . . . . . . . . . . . . . . . Return code 451
        invalid values for time variable
        For instance, you specified mytime as a time variable, and
        mytime contains noninteger values.
```

Often, the error `r(451)` occurs because at least one panel in the dataset has two or more observations that share the same date, dates are not detailed enough to allow these observations to coexist, or both.

In the following toy example, **`xtreg, fe`** (see [XT] **xtreg**) is fit to a short panel dataset ($N > T$) composed of six subsidiaries of the Bank of Alfa that settle their mutual transactions in foreign currencies (values in €2021) at fixed time slots during the first two weeks of November 2021 (table 1):

```
. list bank op_type op_amnt eventdate eventdate2 daily_inc, noobs
```

Table 1. Transactions in foreign currencies

bank	op_type	op_amnt	eventdate	eventdate2	daily_inc
Bank of Alfa 1	Exchange from U.K.£ to €	1000	04/11/2021	04/11/2021 14:32	34887.17
Bank of Alfa 1	Exchange from U.K.£ to €	2000	04/11/2021	04/11/2021 17:20	26688.57
Bank of Alfa 1	Exchange from U.S.$ to €	3000	05/11/2021	05/11/2021 11:20	13664.63
Bank of Alfa 1	Exchange from U.K.£ to €	4000	05/11/2021	05/11/2021 18:36	2855687
Bank of Alfa 1	Exchange from U.K.£ to €	5000	08/11/2021	08/11/2021 10:08	86893.33
Bank of Alfa 2	Exchange from U.S.$ to €	6000	04/11/2021	04/11/2021 14:32	35085.49
Bank of Alfa 2	Exchange from U.S.$ to €	7000	04/11/2021	04/11/2021 17:20	7110509
Bank of Alfa 2	Exchange from U.K.£ to €	8000	05/11/2021	05/11/2021 11:20	32336.79
Bank of Alfa 2	Exchange from U.S.$ to €	9000	05/11/2021	05/11/2021 18:36	55510.32
Bank of Alfa 2	Exchange from U.K.£ to €	10000	08/11/2021	08/11/2021 10:08	87599.10
Bank of Alfa 3	Exchange from U.S.$ to €	2000	04/11/2021	04/11/2021 14:32	20470.95
Bank of Alfa 3	Exchange from U.K.£ to €	4000	04/11/2021	04/11/2021 17:20	89275.87
Bank of Alfa 3	Exchange from U.S.$ to €	6000	05/11/2021	05/11/2021 11:20	58446.58
Bank of Alfa 3	Exchange from U.K.£ to €	8000	05/11/2021	05/11/2021 18:36	36977.91
Bank of Alfa 3	Exchange from U.K.£ to €	10000	08/11/2021	08/11/2021 10:08	85063.09
Bank of Alfa 4	Exchange from U.S.$ to €	12000	04/11/2021	04/11/2021 14:32	39138.19
Bank of Alfa 4	Exchange from U.S.$ to €	14000	04/11/2021	04/11/2021 17:20	11966.13
Bank of Alfa 4	Exchange from U.K.£ to €	16000	05/11/2021	05/11/2021 11:20	75424.34
Bank of Alfa 4	Exchange from U.K.£ to €	18000	05/11/2021	05/11/2021 18:36	69502.34
Bank of Alfa 4	Exchange from U.S.$ to €	20000	08/11/2021	08/11/2021 10:08	68661.52
Bank of Alfa 5	Exchange from U.K.£ to €	3000	04/11/2021	04/11/2021 14:32	93193.46
Bank of Alfa 5	Exchange from U.S.$ to €	4000	04/11/2021	04/11/2021 17:20	45488.82
Bank of Alfa 5	Exchange from U.S.$ to €	5000	05/11/2021	05/11/2021 11:20	6740.11
Bank of Alfa 5	Exchange from U.K.£ to €	6000	05/11/2021	05/11/2021 18:36	33798.89
Bank of Alfa 5	Exchange from U.K.£ to €	7000	08/11/2021	08/11/2021 10:08	97488.48
Bank of Alfa 6	Exchange from U.S.$ to €	12000	04/11/2021	04/11/2021 14:32	72643.84
Bank of Alfa 6	Exchange from U.K.£ to €	14000	04/11/2021	04/11/2021 17:20	4541.51
Bank of Alfa 6	Exchange from U.S.$ to €	16000	05/11/2021	05/11/2021 11:20	74596.66
Bank of Alfa 6	Exchange from U.K.£ to €	18000	05/11/2021	05/11/2021 18:36	49612.59
Bank of Alfa 6	Exchange from U.S.$ to €	20000	08/11/2021	08/11/2021 10:08	71671.62

LEGEND: `daily_inc` = daily income of the bank subsidiary; `op_type` = operation type; `op_amnt` = operation amount.

Transactions are registered via two releases of the same software:

1. an old-fashioned release that accounts only for day/month/year (`eventdate`)[1] and
2. an updated release that also registers hour/minute/second for each transaction (`eventdate2`).

As expected, with the old-fashioned release, Stata warns about repeated dates:

```
. xtset bank eventdate
repeated time values within panel
r(451);
```

The error `r(451)` occurs because `eventdate` shows calendar ties that make it impossible for Stata to sort the dates unambiguously.

Conversely, the software updated release fixes the calendar ties via a more detailed *timevar* (`eventdate2`), and consequently Stata does not issue the error message `r(451)` (table 2):[2]

```
. xtset bank eventdate2
Panel variable: bank (strongly balanced)
Time variable: eventdate2, 04nov2021 14:32:12 to 08nov2021 10:08:01, but with gaps
Delta: .001 seconds
. xtreg op_amnt i.op_type c.daily_inc i.eventdate2, fe
```

1. In fact, Stata allows dates with millisecond precision for this kind of transaction (see [FN] **Date and time functions**).
2. For all the `xtreg` toy examples on bank transactions reported in this tip, the default standard errors (SEs) have been left in because the limited number of groups would have caused the cluster–robust SEs to be potentially misleading (Cameron and Miller 2015).

Table 2. Time of transaction registered via the software updated release (SE)

	op_amnt	
Exchange operation		
Exchange from U.S.$ to €	4.661	
	(414.106)	
Daily income bank subsidiary	0.000	
	(0.008)	
Detailed *timevar*		
04nov2021 17:20:32	1505.253	*
	(559.358)	
05nov2021 11:20:54	3001.388	**
	(529.084)	
05nov2021 18:36:25	4504.244	**
	(578.601)	
08nov2021 10:08:01	5993.361	**
	(586.230)	
Intercept	5984.908	**
	(678.487)	
Number of observations	30	
Number of groups	6.00	
Largest group size	5.00	
F statistic	27.00	
R^2 for within model	0.90	
R^2 for between model	0.34	
R^2 for overall model	0.14	
R^2	0.90	
Adjusted R^2	0.84	
Panel-level standard deviation	5655.23	
Standard deviation of ϵ_{it}	912.85	
ρ	0.97	

** $p < 0.01$, * $p < 0.05$
LEGEND: `op_amnt` = operation amount.

When Stata throws the error `r(451)`, the usual fix is to `xtset` the dataset as in syntax 1. However, this fix comes at the cost of making time-series operators (such as lags and leads) unavailable because they require observations within each panel to be ordered according to *timevar*. Therefore, if time-series operators must be included in the regression equation, the dataset should be `xtset` as in syntax 2.

3 Can timevar still be used as a predictor after error r(451)?

Provided that no variable is differenced, lagged, or led, running `xtreg, fe` as in syntax 1 is perfectly appropriate. It also allows the *timevar* to be plugged in as a categorical predictor in the regression equation despite the error `r(451)` (table 3):

```
. xtset bank eventdate
repeated time values within panel
r(451);
. xtreg op_amnt i.op_type c.daily_inc i.eventdate, fe
```

Table 3. `xtreg, fe` with `i.`*timevar* among predictors despite error `r(451)` after `xtset` (SE)

	`op_amnt`
Exchange operation	
Exchange from U.S.$ to €	−600.853
	(467.742)
Daily income bank subsidiary	−0.009
	(0.010)
Problematic *timevar*	
05nov2021	2922.614 **
	(470.085)
08nov2021	5503.661 **
	(689.394)
Intercept	7477.848 **
	(625.356)
Number of observations	30
Number of groups	6.00
Largest group size	5.00
F statistic	24.01
R^2 for within model	0.83
R^2 for between model	0.54
R^2 for overall model	0.10
R^2	0.83
Adjusted R^2	0.75
Panel-level standard deviation	5777.22
Standard deviation of ϵ_{it}	1136.98
ρ	0.96

** $p < .01$, * $p < .05$
LEGEND: `op_amnt` = operation amount

4 Switching from xtreg, fe to areg when xtset returns error r(451): A good idea?

A tempting work-around for the error `r(451)` is switching from `xtreg, fe` to `areg` (see [R] **areg**) because the latter does not require `xtset`.

Unfortunately, this is not a good idea even in the absence of error `r(451)`, because of the consequences for cluster–robust SE calculation (Cameron and Miller 2015).

Let's expand on this issue using a well-known Stata dataset (table 4):

```
. webuse nlswork
(National Longitudinal Survey of Young Women, 14-24 years old in 1968)
. xtset idcode
Panel variable: idcode (unbalanced)
. xtreg ln_wage c.age##c.age i.year, fe vce(cluster idcode)
. areg ln_wage c.age##c.age i.year, absorb(idcode) vce(cluster idcode)
```

Table 4. `xtreg, fe` versus `areg`: A comparison (cluster–robust SE)

	xtreg, fe		areg	
Age in current year	0.073	**	0.073	**
	(0.014)		(0.015)	
Age in current year # Age in current year	−0.001	**	−0.001	**
	(0.000)		(0.000)	
Interview year				
69	0.065	**	0.065	**
	(0.016)		(0.017)	
70	0.028		0.028	
	(0.026)		(0.029)	
71	0.058		0.058	
	(0.038)		(0.042)	
72	0.051		0.051	
	(0.050)		(0.055)	
73	0.042		0.042	
	(0.062)		(0.068)	
75	0.015		0.015	
	(0.086)		(0.094)	
77	0.034		0.034	
	(0.111)		(0.121)	
78	0.054		0.054	
	(0.123)		(0.135)	
80	0.037		0.037	
	(0.147)		(0.161)	

Continued on next page

	xtreg, fe	areg
Interview year, cont.		
82	0.039	0.039
	(0.172)	(0.188)
83	0.059	0.059
	(0.184)	(0.201)
85	0.104	0.104
	(0.208)	(0.228)
87	0.124	0.124
	(0.233)	(0.255)
88	0.190	0.190
	(0.249)	(0.272)
Intercept	0.394	0.394
	(0.247)	(0.270)
Number of observations	28510	28510
F statistic	79.11	66.04
Number of groups	4710.00	
Largest group size	15.00	
R^2 for within model	0.12	
R^2 for between model	0.11	
R^2 for overall model	0.09	
R^2	0.12	0.67
Adjusted R^2	0.12	0.60
Panel-level standard deviation	0.40	
Standard deviation of ϵ_{it}	0.30	
ρ	0.64	

** $p < 0.01$, * $p < 0.05$

`xtreg, fe` and `areg` produce identical point estimates but different cluster–robust estimates of the variance matrix (Cameron and Miller 2015), because they make different assumptions about whether the number of panels increases with the sample size. While `xtreg, fe` gives back the correct cluster–robust estimates of the variance matrix, `areg` does not, because it uses the wrong degrees-of-freedom correction (Cameron and Miller 2015). This difference, which is particularly apparent when the number of observations per cluster is small, does not hold for default SEs.[3]

3. When the number of observations per cluster is small, the cluster–robust SEs estimated by `areg` should actually be multiplied by the square root of (Cameron and Miller 2015)

$$\{N-(K-1)\}/\{N-G-(K-1)\}$$

N = number of observations, K = number of regressors (intercept included), and G = number of clusters.

5 Leaving out timevar and exploiting the xt commands capabilities: The case of xtgee

The Stata command `xtgee` (see [XT] **xtgee**) fits both linear and nonlinear population-averaged panel-data models via generalized estimating equations (Hardin and Hilbe 2013). Being as flexible as generalized linear models (Deb, Norton, and Manning 2017; Hardin and Hilbe 2018), `xtgee` allows different within-panel correlation structures (via the `corr()` option), various link functions that relate the outcome to the linear index function in the right-hand side of the regression equation (via the `link()` option), and a set of theoretical probability distributions from which the regressand is generated (via the `family()` option). `xtgee`, which does not need a *timevar*, is asymptotically equivalent to `xtreg, re` and `xtreg, mle` (table 5).[4] When panel datasets are balanced, `xtgee` and `xtreg, mle` produce identical results. This equivalence does not hold when panels are unbalanced, because these two Stata commands deal with lack of panel balance differently.

```
. webuse nlswork
(National Longitudinal Survey of Young Women, 14-24 years old in 1968)
. xtset idcode
. xtgee ln_wage grade c.age##c.age, family(gaussian) link(identity)
> corr(exchangeable) nolog
. xtreg ln_wage grade c.age##c.age, re vce(cluster idcode)
. xtreg ln_wage grade c.age##c.age, mle vce(cluster idcode)
```

4. `xtgee` SEs are clustered on *panelvar* by default.

Table 5. xtgee, xtreg, re, and xtreg, mle: A comparison (SE)

	xtgee	xtreg, re	xtreg, mle
Current grade completed	0.080 **	0.080 **	0.080 **
	(0.002)	(0.002)	(0.002)
Age in current year	0.054 **	0.054 **	0.054 **
	(0.003)	(0.004)	(0.004)
Age in current year #	−0.001 **	−0.001 **	−0.001 **
Age in current year	(0.000)	(0.000)	(0.000)
Intercept	−0.369 **	−0.370 **	−0.370 **
	(0.045)	(0.061)	(0.061)
Number of observations	28508	28508	28508
Number of groups	4708.00	4708.00	4708.00
Largest group size	15.00	15.00	15.00
R^2 for within model		0.11	
R^2 for between model		0.32	
R^2 for overall model		0.24	
χ^2	5302.26	3050.74	3058.57
Panel-level standard deviation		0.31	0.30
Standard deviation of ϵ_{it}		0.30	0.30
ρ		0.50	0.49

** $p < 0.01$, * $p < 0.05$

6 Repeated cross-sectional studies and xt commands

In repeated cross-sectional (RCS) studies, a different sample of units per wave is measured on the same set of variables at a defined time point, as in a survey (Lebo and Weber 2015).[5]

Provided that the regressand is continuous, RCS studies are composed of multiple waves of data to be `appended` (see [D] **append**) before running `regress` (see [R] **regress**).

According to the characteristics above, RCS studies fall outside the `xtset` framework.

However, their analysis can benefit from some of the `xt` commands that are frequently used to study panel datasets before running `xt`-related regressions.

5. Often, RCS studies' units are correlated within the same wave and across waves (Lebo and Weber 2015). However, because of the limited number of waves of data, that would cause SEs clustered on `i.year` to be potentially misleading (Lebo and Weber 2015). These issues are not explored in this tip.

A series of one-year RCS data was created by slightly tweaking the `nlswork.dta` file:

```
. webuse nlswork
(National Longitudinal Survey of Young Women, 14-24 years old in 1968)
. sort year
. drop if year>78
. generate cross_sectional_id=_n
. order cross_sectional_id, first
. keep cross_sectional_id ln_wage tenure race not_smsa south year wks_ue
```

The RCS dataset has been `xtset` with *panelvar* only to summarize the continuous variables (table 6):[6]

```
. xtset cross_sectional_id
. xtsum ln_wage tenure wks_ue
```

Table 6. `xtsum` applied to an RCS dataset

Variable		Mean	Std. dev.	Min	Max	Observations
`ln_wage`	overall	1.578669	0.4219723	0.0044871	4.242752	$N = 16094$
	between		0.4219723	0.0044871	4.242752	$n = 16094$
	within		0	1.578669	1.578669	$T = 1$
`tenure`	overall	1.865325	2.081362	0	18.5	$N = 15806$
	between		2.081362	0	18.5	$n = 15806$
	within		0	1.865325	1.865325	$T = 1$
`wks_ue`	overall	2.371952	6.861621	0	56	$N = 15709$
	between		6.861621	0	56	$n = 15709$
	within		0	2.371952	2.371952	$T = 1$

As expected, the overall and between standard deviations overlap (because $N = n$), whereas the within one is zero (because $T = 1$).

The `xtsum` outcome table mirrors the wide range of the continuous variables.

In addition, RCS datasets allow time-fixed effects to account for variations over time (table 7):[7,8]

```
. regress ln_wage c.tenure##c.tenure i.race i.not_smsa i.south i.year wks_ue,
> vce(robust)
```

6. Note that, unlike `xtsum` (see [XT] **xtsum**), `xtdescribe` (see [XT] **xtdescribe**) requires the dataset to be `xtset` with a *timevar* too.
7. `_robust` (see [P] **_robust**) SEs were imposed because of heteroskedasticity of the residual distribution checked via `estat hettest` (Prob $> \chi^2 = 0.0399$).
8. The joint statistical significance of `i.year` was tested via `testparm`: $F(8, 15403) = 16.56$; Prob $> F = 0.0000$, whereas the correct specification of the functional form of the regressand was confirmed via `linktest` ([R] **linktest**): `_hatsq` $P > |t| = 0.388$ (`linktest` returns prediction squared as `_hatsq`).

Table 7. Ordinary least squares (OLS) on an RCS dataset (robust SE)

	RCS_OLS	
Job tenure in years	0.114	**
	(0.004)	
Job tenure in years # Job tenure in years	−0.007	**
	(0.000)	
Race		
Black	−0.100	**
	(0.007)	
Other	−0.011	
	(0.027)	
1 if not standard metropolitan statistical area		
1	−0.189	**
	(0.007)	
1 if south		
1	−0.133	**
	(0.006)	
Interview year		
69	0.049	**
	(0.014)	
70	−0.003	
	(0.014)	
71	0.025	
	(0.013)	
72	0.027	
	(0.014)	
73	0.029	*
	(0.014)	
75	0.021	
	(0.013)	
77	0.085	**
	(0.014)	
78	0.112	**
	(0.014)	
Weeks unemployed last year	−0.003	**
	(0.001)	
Intercept	1.524	**
	(0.011)	
Number of observations	15419	
F statistic	316.91	
R^2	0.23	
Adjusted R^2	0.23	

** $p < .01$, * $p < .05$

7 Conclusion

This tip started from the evidence of frequent complaints about the error `r(451)` posted on the Stata forum and then expanded to other `xt`-related issues.

`xtset` has two dimensions to be addressed: the cross-sectional one (*panelvar*), which is mandatory because it tells Stata that the researcher is dealing with a panel dataset, and an optional one, that is, the time-series dimension (*timevar*).

Therefore, how to `xtset` the panel dataset is strictly related to the study goals.

Unlike the `xtabond` case (see [XT] **xtabond**), various panel-data commands that provide useful information without the need of a time variable, for example, `xtsum` for RCS studies, can give the researcher more information on standard deviation than `summarize`.

8 Acknowledgment

I thank Nicholas J. Cox for his constructive comments.

References

Cameron, A. C., and D. L. Miller. 2015. A practitioner's guide to cluster–robust inference. *Journal of Human Resources* 50: 317–372. https://doi.org/10.3368/jhr.50.2.317.

Cameron, A. C., and P. K. Trivedi. 2005. *Microeconometrics: Methods and Applications.* New York: Cambridge University Press.

———. 2022. *Microeconometrics Using Stata.* 2nd ed. College Station, TX: Stata Press.

Deb, P., E. C. Norton, and W. G. Manning. 2017. *Health Econometrics Using Stata.* College Station, TX: Stata Press.

Hardin, J. W., and J. M. Hilbe. 2013. *Generalized Estimating Equations.* 2nd ed. Boca Raton, FL: CRC Press.

———. 2018. *Generalized Linear Models and Extensions.* 4th ed. College Station, TX: Stata Press.

Lebo, M. J., and C. Weber. 2015. An effective approach to the repeated cross-sectional design. *American Journal of Political Science* 59: 242–258. https://doi.org/10.1111/ajps.12095.

Wooldridge, J. M. 2020. *Introductory Econometrics: A Modern Approach.* 7th ed. Boston: Cengage Learning.

The Stata Journal (2023)
23, Number 1, pp. 293–297 DOI: 10.1177/1536867X231162009

Stata tip 151: Puzzling out some logical operators

Nicholas J. Cox
Department of Geography
Durham University
Durham, U.K.
n.j.cox@durham.ac.uk

The logical operators `&` ("and") and `|` ("or") can sometimes be tricky in statistical software such as Stata. They are extremely useful, so you need to understand thoroughly how they work. Any trickiness arises mostly in translating from ordinary language to a statistical computer language. Here I survey various common confusions and explain what to do instead.

`auto.dta` in Stata will serve fine as a sandbox.

```
. sysuse auto
(1978 automobile data)
```

1 What is wrong, and why

The repair record variable `rep78` in `auto.dta` takes on values 1 (poor) to 5 (best) and also missing. You can see that with a simple tabulation:

```
. tabulate rep78, missing
```

Repair record 1978	Freq.	Percent	Cum.
1	2	2.70	2.70
2	8	10.81	13.51
3	30	40.54	54.05
4	18	24.32	78.38
5	11	14.86	93.24
.	5	6.76	100.00
Total	74	100.00	

Let's see which cars have repair record 1. You need to use the operator `==` when testing for equality. If this point is new to you, please consult `help operators`.

```
. list make rep78 if rep78 == 1
```

	make	rep78
40.	Olds Starfire	1
48.	Pont. Firebird	1

Two cars are shown, as promised by the previous table. Now consider this syntax, which is an attempt to also get those cars with `rep78==2`:

```
. list make rep78 if rep78 == 1 & 2

     +----------------------+
     | make           rep78 |
     |----------------------|
 40. | Olds Starfire      1 |
 48. | Pont. Firebird     1 |
     +----------------------+
```

We get the same cars. Where are the 8 cars with value 2? The command was legal but wrong from our point of view. A legal command is one that runs without an error message, but evidently being legal does not make a command right for us. Now suppose you report your problem to a friend, who explains that you need the "or" operator there, not the "and" operator. It is impossible for a value of a Stata variable to be both 1 and 2 in the same observation. You do want those observations that are 1 AND those observations that are 2. For Stata, that means selecting those observations for which there is value 1 on `rep78` OR for which there is value 2 on `rep78`.

Emphasizing operators by using uppercase (for example, AND, OR) is a practice I learned from John Tukey's writing (for example, Tukey [1977]). As many people have a synesthetic sense that using uppercase is SHOUTING, it is best done sparingly. This is no more than presentation: AND and OR are assumed to behave exactly like `&` and `|` otherwise.

Suppose further that your friend does not spell out syntax, so you try

```
. list make rep78 if rep78 == 1 | 2
```

However, now those variables are listed for all 74 observations in the dataset. (The lengthy listing is not reproduced here, but you can check for yourself.) So, that command too is legal but wrong. Sooner or later—say, your friend is more explicit, or you look at some documentation—you reason or muddle your way toward

```
. list make rep78 if rep78 == 1 | rep78 == 2

     +---------------------------+
     | make                rep78 |
     |---------------------------|
 12. | Cad. Eldorado           2 |
 17. | Chev. Monte Carlo       2 |
 18. | Chev. Monza             2 |
 21. | Dodge Diplomat          2 |
 22. | Dodge Magnum            2 |
     |---------------------------|
 23. | Dodge St. Regis         2 |
 40. | Olds Starfire           1 |
 46. | Plym. Volare            2 |
 48. | Pont. Firebird          1 |
 52. | Pont. Sunbird           2 |
     +---------------------------+
```

Eventually, you got what you wanted, but we now need to explain exactly why those earlier guesses are not right.

The first principle here is that an `if` qualifier selects observations if the stated condition is true. That was easy to specify when the condition was just `rep78 == 1`, but what was happening with a compound condition such as `rep78 == 1 & 2`—where now two logical operators are in sight? The answer lies in precedence of operators, namely, which operator is used first in evaluation. Stata's precedence rules are documented at `help operators`. You do not need to learn the order in which operators are used in evaluation. All that I have found important is knowing how to look up the order and knowing to try to use parentheses, `( )`, to insist on your intended meaning.

By the way, I recommend the terminology whereby `( )`, `[ ]`, and `{ }` are called in turn "parentheses", "brackets", and "braces". In turn, extra adjectives "round", "square", and "curly" are, according to taste, either redundant or revealing. For many other names, see Raymond (1996). Books on punctuation range from splenetic to scholarly: Houston (2013) and especially Parkes (1993) are nearer the latter.

The answer here is that `==` is used before `&`. So, faced with the condition

```
if rep78 == 1 & 2
```

Stata parses it as if it were

```
if (rep78 == 1) & 2
```

Just as in school mathematics, what is inside parentheses is treated first in evaluation. We now need to know Stata's rule that expressions evaluating to zero (0) are false while expressions evaluating to any number other than zero are true. This last detail may be new to you, especially if you are more familiar with the vital implication, which is very widely useful, that 0 means false and 1 means true (Cox 2005, 2016; Cox and Schechter 2019). Imagine just the single condition

```
if 2
```

Clearly, 2 is not 0, so 2 counts as true—always, meaning for every observation. Stata does look at every observation and asks, with this syntax, whether 2 is true, given the information in this observation, to which the answer is always yes. Even though it seems unlikely to be something you would write on purpose, the syntax is legal and has meaning. For completeness, note that code such as `if 2` is legal and has meaning also even if there are no data in memory.

So, `if 2` is always true, but the compound condition `if rep78 == 1 & 2` is true only when both conditions are true. That restriction narrows the scope to observations `if rep78 == 1`, as already observed.

Operator precedence also means that `==` is used in evaluation before `|`, so `rep78 == 1|2` is evaluated as if it were `(rep78 == 1)|2`. The compound condition is true if either condition is true, and as already observed, 2 is always true; and so the entire condition is always true, with the consequence reported earlier that all observations are used whenever the condition is `rep78 == 1 | 2`.

It should now seem clear that compound conditions such as `if rep78 == 1 | 2 | 3 | 4 | 5` do not offer a terse and generally applicable syntax for multiple conditions acting at the same time. But there remains much scope for small puzzles. Thus, the last-stated condition almost does what may have been intended (it catches missing values too).

The `foreign` indicator variable takes on values 0 and 1 only in `auto.dta`. It follows that `if foreign == 0 | 1` does what may have been intended—catch observations with values of both 0 and 1 on `foreign`—but by accident because the true condition `1` catches all observations. Conversely, the condition `if foreign == 1 | 0` is not at all equivalent and will not catch any values of 0, because 0 counts as false. Otherwise put,

```
if foreign == 1 OR 0
```

and

```
if foreign == 0 OR 1
```

are not equivalent pseudocode.

2 Other ways to get it right

Earlier mentions may have left the impression that slow but sure is the only successful tactic in specifying compound conditions. The point of this section is to emphasize other syntax and other tactics. See also Cox (2006, 2011).

`inlist(rep78, 1, 2)` is another way to write `if rep78 == 1 | rep78 == 2`, and its appeal grows with the number of possible values given in the list, here just 1 and 2, but in many problems a longer list. See the help for `inlist()` for current limits on the number of arguments.

Mention must be made of the useful twist that `inlist(1, a, b, c, d, e)`, mentally expanded as `1 == a | 1 == b | 1 == c | 1 == d | 1 == e`, is thus a way of checking that any of `a`, `b`, `c`, `d`, `e` is 1 just as surely as `1 == a` is equivalent to `a == 1`.

Because `rep78` takes only integer values, `inrange(rep78, 1, 2)` is in practice equivalent, as is `rep78 <= 2` or `rep78 < 3`. There are differences in principle. In the first case, that difference is the possibility of noninteger values in the stated range. In the second case, that kind of problem could apply, together with the possibility of values below 1.

It is always worth flagging that numeric missing values all count as nonzero and hence as true.

3 Strings cannot be true or false

This section separates off a warning that because string values are not numeric, they cannot themselves be true or false. Thus, suppose that we wanted all the available information on the two cars identified by the first line of code in section 2.

```
. list if make == "Olds Starfire" | "Pont. Firebird"
type mismatch
r(109);
```

It should now not be surprising that this code is not what is wanted, but in this case, the code is illegal and so triggers an error message. The whole of `make == "Olds Starfire"` is a true or false expression evaluated as 1 or 0, but `"Pont. Firebird"` as a bare string cannot be true or false; hence, the error shown.

```
. list if make == "Olds Starfire" | make == "Pont. Firebird"
```

and

```
. list if inlist(make, "Olds Starfire", "Pont. Firebird")
```

are fine ways to issue the instruction.

References

Cox, N. J. 2005. FAQ: What is true or false in Stata? https://www.stata.com/support/faqs/data-management/true-and-false/.

———. 2006. Stata tip 39: In a list or out? In a range or out? *Stata Journal* 6: 593–595. https://doi.org/10.1177/1536867X0600600413.

———. 2011. Speaking Stata: Fun and fluency with functions. *Stata Journal* 11: 460–471. https://doi.org/10.1177/1536867X1101100308.

———. 2016. Speaking Stata: Truth, falsity, indication, and negation. *Stata Journal* 16: 229–236. https://doi.org/10.1177/1536867X1601600117.

Cox, N. J., and C. B. Schechter. 2019. Speaking Stata: How best to generate indicator or dummy variables. *Stata Journal* 19: 246–259. https://doi.org/10.1177/1536867X19830921.

Houston, K. 2013. *Shady Characters: Ampersands, Interrobangs and other Typographic Curiosities*. London: Particular Books.

Parkes, M. B. 1993. *Pause and Effect: Punctuation in the West*. Berkeley, CA: University of California Press.

Raymond, E. S. 1996. *The New Hacker's Dictionary*. 3rd ed. Cambridge, MA: MIT Press.

Tukey, J. W. 1977. *Exploratory Data Analysis*. Reading, MA: Addison–Wesley.

178 DOI: 10.1177/1536867X231175349

The Stata Journal (2023)
23, Number 2, pp. 589–594

Stata tip 152: if and if: When to use the if qualifier and when to use the if command

Nicholas J. Cox
Department of Geography
Durham University
Durham, U.K.
n.j.cox@durham.ac.uk

Clyde B. Schechter
Albert Einstein College of Medicine
Bronx, NY
clyde.schechter@einsteinmed.edu

1 Introduction

Stata has an `if` qualifier and an `if` command. Here we discuss generally when you should use either and specifically flag a common pitfall in using the `if` command. In a nutshell, the pitfall arises from confusing the two constructs: the `if` command does not loop over the data but, at most, looks in the first observation of a dataset. There has long been a StataCorp FAQ on this topic (Wernow 2005), but we and others have usually tried to explain matters otherwise. This tip is intended as a more durable version of the story that should be easier to find than occasional Statalist postings that are vivid when read but hard to find later.

2 The if qualifier

The `if` qualifier is met by most users early in their Stata experience. Its purpose is to select observations (cases, records, or rows in the dataset) for some action. Thus, you could run the following commands to read in a dataset and first `summarize` a variable and then `summarize` that variable again for a subset of observations. Here we suppress the results, but if you are new to Stata and unfamiliar with `summarize`, it would be worth your time to run the code yourself to find out about a valuable command.

```
. sysuse auto
. summarize mpg
. summarize mpg if foreign == 1
```

When the `if` qualifier is used (or, in other words, when an `if` condition is specified), Stata tests the expression given—here `foreign == 1`—in each observation to see whether it is satisfied (is true) in that observation. Observations for which the expression is true are selected for the action. In this example, `foreign` is an indicator variable that is 1 if a car is foreign (made outside the United States) and 0 if a car is domestic (made inside the United States). The operator `==` tests for equality, noting that in Stata the `=` operator typically indicates assignment of a value or values, say, to a variable. Out of 74 cars, 22 qualify as being foreign, so their observations will be `summarize`d for the variable `mpg`.

Stata follows a very widely used convention, running across statistics, mathematics, and computing, that in logical tests, a value of 1 means true and a value of 0 means false. In fact, Stata's rule is more general: Any numeric value that is not 0 means true, while only the numeric value 0 means false. Watch out with missing values because any numeric value that represents missing (whether system missing, `.`, or extended missing values from `.a` to `.z`) is certainly not 0 and so yields true in a logical test.

Logical tests in Stata take two forms. First, and more commonly, some logical operator is used in an expression. Tests for equality, using the `==` operator, may be what you need; otherwise, some test for inequality may be needed. See the help for operators to see the complete list. Thus, in `auto.dta` you could select cars with high `mpg` by, say, `mpg > 25`. Logical tests can combine two or more conditions, but even so the keyword `if` appears only once in any comparison.

Second, you can ask Stata to look inside a numeric variable and check whether its values are 0 or not. In `auto.dta`, `foreign` is only ever 1 or 0 and never missing. So the test `if foreign` is precisely the same test in practice as `if foreign == 1`. Presented with `if foreign`, Stata looks inside the variable and selects those observations for which it is not 0, which in practice is the same subset of observations as those for which the condition `if foreign == 1` is true.

There are positive and negative sides to this flexibility. The positive side is that we can write Stata code that may appeal to readers as idiomatic in their own language and in Stata too. "Let's focus on the cars that are foreign" becomes the condition `if foreign`. Such coding works best if you follow a convention, which we strongly recommend, of naming an indicator variable for the condition coded as 1. That is precisely what the developers of Stata did at the very beginning when coding up the auto data.

The negative side is that the inclusiveness here could bite if there are nonzero values that the condition `if foreign` would catch too, even though that is not what you intend. As said, nonzero values include any numeric missing values. So you might well prefer to be safe rather than succinct and always spell out, say, `if foreign == 1`.

For more on truth and falsity in Stata, see Cox (2005, 2016). For more on indicator variables, see Cox and Schechter (2019), especially if you have been thinking "Don't you mean dummy variables?" (Yes, we do.)

3 The if command

The previous section may have strengthened your understanding of the `if` qualifier, say, by spelling out some nuances. At this point in the story, the most important detail about the `if` command is that it is emphatically not a way to do the same thing differently. Oddly, or otherwise, a misunderstanding that the two are equivalent (or at least overlap in what they do) seems to arise most often with people new to Stata who are accustomed to programming in some other language. Such programmers may guess or hope that Stata's `if` command is similar to, or an extension of, what they know already.

Whatever the explanation, constructs using `if` or some equivalent keyword have been present in many programming languages over several decades. Examples can be found in Sammet (1969), Kernighan and Plauger (1978), and Bal and Grune (1994).

We will pursue this negative theme before turning to when and why the `if` command is appropriate or useful. Otherwise, there would be no point to including it within Stata.

Any puzzlement is intensified whenever Stata allows use of the `if` command in a way that seems equivalent to use of the `if` qualifier. It then gives results that occasionally are what you want but more often just seem bizarre. As examples, consider these two statements and their results:

```
. if foreign == 1 summarize mpg
. if foreign == 0 summarize mpg
    Variable |        Obs        Mean    Std. dev.       Min        Max
-------------+---------------------------------------------------------
         mpg |         74     21.2973    5.785503         12         41
```

Stata complains about neither statement, so each is perfectly legal. But you might even wonder whether you have unearthed a bug. The first statement yields no results, whereas we already know that there are observations for which `foreign == 1`. Other way round, the second statement yields results, but if you look carefully, you will see that the results are for the entire dataset and so include both foreign and domestic cars.

The explanation is immediate given one extra piece of information. When an `if` command refers to a variable (or variables) in the dataset, Stata looks only in the first observation. It is exactly as if you wrote `if foreign[1] == 1` or `if foreign[1] == 0`. It so happens that the first statement is false and the second statement is true, as can be checked independently by looking at the data with, say, `list in 1` or `edit in 1` or `display foreign[1]`. Because the first statement was false, Stata did not execute the next command, `summarize mpg`. Because the second statement was true, Stata did execute the (same) next command. In both cases, the subset of observations specified was not part of the syntax for the next command.

We could make that plainer by writing the same syntax using curly brackets or braces:

```
if foreign[1] == 0 {
    summarize mpg
}
```

Backing up slightly: Here a so-called subscript such as `[1]` attached to a variable name indicates an observation number, so in another example `foreign[42]` would be the value of `foreign` in observation 42. We say "subscript" as an allusion to mathematical notation such as y_1 or y_{42}, but naturally writing *sub scriptum*, below the line, is not strictly possible in Stata.

A more general point to emphasize is that there is no sense in Stata in which the `if` command iterates or loops over the observations in the dataset. (Here we are assuming that there are data in memory; it is perfectly possible to use Stata with no variables in memory, and you may wish to think through what could be done depending on what else is allowed.) Positively put, the `if` command makes one and only one decision, depending on whether the condition specified is true.

The `if` command is very widely used within do-files and within programs, including within programs that define other commands.

There are many examples within Stata programs. Options are typically implemented in this way. In many commands, there are optional choices, either for extra actions or to vary some action from the default. Inside the command code, there is typically a switch for each option whereby different code is executed. The `summarize` command has options, such as `meanonly` (to do less than the default) or `detail` (to do more). That command is built in, so users may not see the internal code, but very many commands are implemented through ado-code, so much of or all the code is visible. If you are curious, you can look inside ado-code with, say,

```
. viewsource tabstat.ado
```

and you will immediately see a series of switches all using the `if` command to set up calculations according to whatever a user did (or did not) specify when issuing the `tabstat` command.

Another common sequence within ado-code is something like this.

```
. marksample touse
. count if `touse´
  74
. if r(N) == 0 error 2000
```

Here `marksample` has the job of creating a temporary indicator variable `` `touse' `` that is 1 when observations are to be used and 0 otherwise. (If the name `touse` looks odd to you, think "to use".) Exclusions arise for one of two reasons: whenever missing values make the use of observations impossible or whenever an `if` qualifier (there it is again) or an `in` qualifier excludes observations by implication. We then `count` the observations to be used. The result is left in `r(N)`. If that result is 0, then there are no observations to use, which here and usually is regarded as an error. If, as it were, no news is good news, such as when we are checking for something bad but fail to find it, then the syntax would be different. We might well condition on, say, `r(N) > 0`.

There are other vital differences between the `if` qualifier and the `if` command, beyond the cosmetic (but still crucial) difference that the first follows and the second precedes associated code.

The `if` command can be associated with code following `else` to indicate what should be done if the condition specified is false. Indeed, a more or less complicated series of branching decisions may be needed depending on a menu of possible choices. Again, if you are curious, look at the results of

```
. viewsource duplicates.ado
```

which show a series of branches aimed at identifying the subcommand that a user specified after the command itself, such as `duplicates report` or `duplicates list`.

Lest you think that the `if` command is primarily of interest to Stata programmers, let's look at an example of its use in a common situation that arises in data analysis. Suppose you want to analyze some panel data, performing some specific calculations separately in each panel but only in those panels that offer a minimum sample size. Here we assume for simplicity that firms have distinct numeric identifiers. The code in your do-file might look like this:

```
generate abnormal_return = .
levelsof firm, local(firms)
foreach f of local firms {
    count if firm == `f'                                // N.B. if qualifier
    if r(N) >= 30 {                                     // N.B. if command
        regress return market_return if firm == `f'     // if qualifier
        predict resid, resid
        replace abnormal_return = resid if firm == `f' // if qualifier
        drop resid
    }
}
```

Notice that both the `if` command and the `if` qualifier are used in this code, with very different effects. The `if` qualifier applies only to the single command in which it appears, and it restricts those commands to the observations for which `` firm == `f' ``. The `if` command appears only once in the code, but it controls execution of the following four commands; they are executed only if the result of the preceding `count` command is at least 30. Note, in particular, that this `if` command does not examine any observations in the data in memory: it refers only to the result returned by the preceding `count` command. Note also the use of curly braces to apply the single `if` command to an entire block of commands. Those four commands are all executed, or none are, depending on the available sample size for the firm.

You may be thinking of refinements, such as counting observations with nonmissing values, because observations with missing values are of no use for any regression. You may also know of community-contributed commands in this area, but discussing those is beyond our scope here. Even if you have access to such commands, understanding the principles in this last example is valuable in many contexts.

References

Bal, H. E., and D. Grune. 1994. *Programming Language Essentials*. Wokingham: Addison–Wesley.

Cox, N. J. 2005. FAQ: What is true or false in Stata? https://www.stata.com/support/faqs/data-management/true-and-false/.

———. 2016. Speaking Stata: Truth, falsity, indication, and negation. *Stata Journal* 16: 229–236. https://doi.org/10.1177/1536867X1601600117.

Cox, N. J., and C. B. Schechter. 2019. Speaking Stata: How best to generate indicator or dummy variables. *Stata Journal* 19: 246–259. https://doi.org/10.1177/1536867X19830921.

Kernighan, B. W., and P. J. Plauger. 1978. *The Elements of Programming Style*. New York: McGraw–Hill.

Sammet, J. E. 1969. *Programming Languages: History and Fundamentals*. Englewood Cliffs, NJ: Prentice-Hall.

Wernow, J. 2005. FAQ: I have an if or while command in my program that only seems to evaluate the first observation. What's going on? http://www.stata.com/support/faqs/programming/if-command-versus-if-qualifier/.